HENRY SEEBOHM

CLASSIFICATION

OF

BIRDS

AN ATTEMPT TO DIAGNOSE THE SUBCLASSES, ORDERS, SUBORDERS AND SOME OF THE FAMILIES OF EXISTING BIRDS

Elibron Classics
www.elibron.com

Elibron Classics series.

ISBN 1-4021-5925-0 (paperback)
ISBN 1-4021-0610-6 (hardcover)

This Elibron Classics Replica Edition is an unabridged facsimile of the edition published in 1890 by R. H. Porter, London.

CLASSIFICATION

BIRDS;

AN ATTEMPT TO DIAGNOSE THE SUBCLASSES, ORDERS, SUBORDERS, AND SOME OF THE FAMILIES OF EXISTING BIRDS.

BY

HENRY SEEBOHM.

LONDON:
R. H. PORTER, 18 PRINCES STREET, CAVENDISH SQUARE, W.
1890.

ALERE FLAMMAM.

PRINTED BY TAYLOR AND FRANCIS,
RED LION COURT, FLEET STREET.

INTRODUCTION.

The following is an attempt to classify existing Birds in groups which are capable of being diagnosed. I have divided the Class Aves into six Subclasses, fourteen Orders, and thirty-six Suborders. These Subclasses, Orders, and Suborders are diagnosed by a series of characters *each* of which is believed to be found in *every* species contained in the group diagnosed, and the combination of which is believed not to occur in any species outside that group. In order to show that each combination is diagnostic, a list of the thirty-six suborders is attached to each diagnosis, and a star is placed under each suborder in which every species it contains is excluded by the character referred to by a letter at the side of the column in which the star is placed. In most cases a few additional characters are given to strengthen the diagnosis. In selecting characters I have been careful to choose those which are supposed to denote affinity—in other words, to be due to inheritance from common ancestors; and as far as possible to avoid such as only denote analogy, that is to say such as are only instances where like causes have produced like effects. It is, however, very difficult to tell which characters have been inherited and which have been independently acquired. In many cases it is only possible to guess, in others it is absolutely impossible to form any opinion at all. It may perhaps be fairly assumed that the more complicated a character is, the less chance is there that it can have been independently acquired by two groups. It may also be taken for granted that it is very much more difficult even for a simple character to be independently acquired than to be independently lost.

It must be remembered that, so far as the individual is concerned, every bird inherited *all* its characters, whether they may be adapted to its present needs or not, but that *all* characters were originally acquired either by the species or its ancestors because they were originally adapted to its needs.

If the first great difficulty of classification be to find out in each case whether a given character denotes affinity or only analogy, the second great difficulty is to ascertain the relative value of the different characters. Unfortunately no rule can be laid down; but there are certain laws which govern the operation of classification by diagnosis. The characters of each group must belong to every species in that group, otherwise it is obvious that artificial groups could be made to any extent. On the other hand, half the species in an outside group may be excluded by one character, and the other half by another character. For example, in diagnosing the Pico-Passeres the absence of the ambiens muscle excludes part of the Columbæ, and the absence of the accessory femoro-caudal muscle excludes the rest. In diagnosing the Columbæ the presence of basipterygoid processes excludes all the Pico-Passeres except the Trogones, and the schizorhinal character of the nasals excludes all the Pico-Passeres except a few Passeres. This is perfectly legitimate, but the converse would not be legitimate. To include amongst the characters of the Columbæ ambiens *or* accessory femoro-caudal muscles present, or to include amongst the characters of the Pico-Passeres nasals not schizorhinal unless accompanied by the absence of basipterygoid processes, though perfectly true, is not admissible, as it would introduce a practice that would make the combination of widely separated groups into an unnatural group perfectly easy.

By strictly obeying this law the number of possible classifications becomes greatly restricted, and the danger of forming unnatural groups is greatly diminished.

The breaking-down of a character in one group is no reason why it should not be employed in groups where it does not break down. For example, the fact that some of the Tubinares possess basipterygoids whilst others are without those processes is no bar to the presence of basipterygoid processes in the Limicolæ being used as a character to exclude the Gaviæ.

The student who attempts to diagnose a group will probably begin by being very saucy in the selection of his characters, and will no doubt arrogantly reject many which are in his opinion too trivial. He will most likely end in gratefully accepting an apparently slight character if it will only run through all the species in the group he is trying to diagnose. He will discover that Nature has decided for him which characters are important and which are trivial, and in which groups they are important and in which groups they are trivial. It may almost be accepted as an axiom that every character breaks down somewhere. The range of variation in Birds is so small that there are scarcely any characters which are not passing through the process of being acquired or lost, as the case may be, in one of the thirty or forty suborders into which birds may be naturally divided.

Fossil birds throw little or no light on the relative age of the various characters.

In the first place fossil birds are very rare ; in the second place they tell us nothing beyond a few facts concerning their osteology ; and in the third place the few facts they do tell us only make the darkness more obscure. The oldest fossil birds, of which numerous and nearly perfect examples have been found, were discovered in the Cretaceous beds of North America. Even at that remote period some birds had lost the keel to the sternum and acquired heterocœlous dorsal vertebræ. So far as we know the oldest birds had teeth, amphicœlous or doubly concave vertebræ, powerful wings, a keel to the sternum, four toes (three in front and one behind), webbed feet, a schizognathous palate, basipterygoid processes, and many other characters too numerous to mention. The geological record is, however, so imperfect that many of these hypotheses are little more than guesses.

Once for all it must be noted that any attempt to bring all fossil birds into the same system of classification as those now living is bound to fail. Between every two closely-allied groups of existing birds there must have been birds now extinct, the common ancestors of both, most probably differing from both, and partly resembling both, and incapable of being classified with either. To encumber the classification of existing birds with a few scattered links in endless chains of intergrading forms can only create confusion. The classification of fossil birds is a most interesting inquiry, and might be called the study of a vertical section of the bird-life which has existed in past geological ages. The classification of existing birds is the study of a horizontal section of the great bird mass of the world, and ought to form a different and distinct system confined to the horizon of the present time.

If birds have all been derived from a common stock, which can scarcely be doubted, every species is in one sense of the same age ; but some have developed slowly, possibly where the conditions of life have been easy, whilst others have developed rapidly where the struggle for existence has been more severe. The various races of man are presumably all descended from one family: but "Quashy," thanks to his "pumpkin," has had a very easy time of it in tropical and semitropical Africa, and has consequently lagged behind wofully in the race ; whilst John Bull has had a hard fight for existence, especially since he was crowded in his tight little island, and in the same time has made rapid strides. The struggle for existence increases the pace of development, if it be not too severe. Where the fight is too hard, as it has been for the Esquimaux on his arctic tundra, the fittest only just survive, or a less highly organized race is most adapted to struggle with such difficulties, and consequently the changes gradually introduced by natural selection are in a retrograde direction.

It is impossible in a work of this kind, where so many facts have to be ascertained and recorded, and so many generalizations have to be made, entirely to avoid blunders. In many cases it is impossible to procure the material necessary to verify the statements of others, but in all doubtful points I have endeavoured to do so. Much work remains to be done before the families of the Passeres, the Columbæ, the Psittaci, and a few other suborders can be satisfactorily diagnosed ; but in the meantime I have ventured to publish my imperfect schemes in the hope that they may induce others to help to fill up the gaps and correct the mistakes, and if possible decide which of the two schemes presented for criticism appears to be the most natural.

The Class AVES may be subdivided as follows :—

SUBORDERS.	ORDERS.	SUBCLASSES.
1. PASSERES.	PICO-PASSERES.	PASSERIFORMES.
2. EURYLÆMI.	PICO-PASSERES.	PASSERIFORMES.
3. TROCHILI.	PICO-PASSERES.	PASSERIFORMES.
4. SCANSORES.	PICO-PASSERES.	PASSERIFORMES.
5. UPUPÆ.	PICO-PASSERES.	PASSERIFORMES.
6. TROGONES.	PICO-PASSERES.	PASSERIFORMES.
7. COLUMBÆ.	COLUMBÆ.	PASSERIFORMES.
8. MUSOPHAGI.	COCCYGES.	PASSERIFORMES.
9. CUCULI.	COCCYGES.	PASSERIFORMES.
10. SERPENTARII.	RAPTORES.	FALCONIFORMES.
11. ACCIPITRES.	RAPTORES.	FALCONIFORMES.
12. STRIGES.	RAPTORES.	FALCONIFORMES.
13. PSITTACI.	PSITTACI.	FALCONIFORMES.
14. HALCYONES.	PICARIÆ.	CORACIIFORMES.
15. CORACIÆ.	PICARIÆ.	CORACIIFORMES.
16. BUCEROTES.	PICARIÆ.	CORACIIFORMES.
17. CATHARTES.	CATHARTES.	CORACIIFORMES.
18. STEGANOPODES.	PELECANO-HERODIONES.	ANSERIFORMES.
19. HERODIONES.	PELECANO-HERODIONES.	ANSERIFORMES.
20. PLATALEÆ.	PELECANO-HERODIONES.	ANSERIFORMES.
21. PHŒNICOPTERI.	PELECANO-HERODIONES. / LAMELLIROSTRES.	ANSERIFORMES.
22. ANSERES.	LAMELLIROSTRES.	ANSERIFORMES.
23. PALAMEDEÆ.	LAMELLIROSTRES.	ANSERIFORMES.
24. TUBINARES.	TUBINARES.	GALLIFORMES.
25. IMPENNES.	IMPENNES.	GALLIFORMES.
26. GAVIÆ.	GALLO-GRALLÆ.	GALLIFORMES.
27. LIMICOLÆ.	GALLO-GRALLÆ.	GALLIFORMES.
28. GRALLÆ.	GALLO-GRALLÆ.	GALLIFORMES.
29. FULICARIÆ.	GALLO-GRALLÆ.	GALLIFORMES.
30. PYGOPODES.	GALLO-GRALLÆ.	GALLIFORMES.
31. GALLINÆ.	GALLO-GRALLÆ.	GALLIFORMES.
32. CRYPTURI.	GALLO-GRALLÆ.	GALLIFORMES.
33. APTERYGES.	APTERYGES.	STRUTHIONIFORMES.
34. RHEÆ.	RATITÆ.	STRUTHIONIFORMES.
35. CASUARII.	RATITÆ.	STRUTHIONIFORMES.
36. STRUTHIONES.	RATITÆ.	STRUTHIONIFORMES.

ALTERNATIVE SCHEME.

It is quite possible that in the foregoing scheme too much importance has been given to the arrangement of the palatine bones and the condition of the feather-tracts of the neck. In the following alternative scheme the Orders and Suborders remain the same, except that their sequence is slightly altered. Three of the Subclasses also remain the same—the Passeriformes, the Coraciiformes, and the Struthioniformes. The other three—the Falconiformes, the Anseriformes, and the Galliformes—are rearranged in two subclasses, the Ciconiiformes and the Galliformes. The arrangement of the palatine bones and the condition of the feather-tracts of the neck are omitted from the diagnoses, but the condition of the young at birth and the disposition of the deep plantar tendons are retained.

The Ciconiiformes may be diagnosed as follows:—**A.** Young born helpless and requiring to be fed by their parents for many days. (This excludes the Galliformes and the Struthioniformes.) **B.** Young passing through a downy stage before acquiring feathers. (This excludes the Passeriformes.) **C.** *Flexor perforans digitorum* never leading to hallux. (This excludes the Coraciiformes.)

The Galliformes as rearranged may be diagnosed as follows:—**A.** Young born able to run in a few hours. (This excludes the Passeriformes, the Coraciiformes, and the Ciconiiformes.) **B.** Keel of sternum well developed. (This excludes the Struthioniformes.)

In many respects the alternative scheme arranges the Orders and Suborders more satisfactorily than the original one. The choice between them may be narrowed down to one or two simple questions:—Are the Tubinares and Impennes, in spite of their schizognathous palates, more nearly allied to the desmognathous Pelecano-Herodiones, because they agree with them in having the young born helpless and requiring to be fed by their parents for many days, than they are to the schizognathous Gallo-Grallæ, which differ from them in having the young born able to run and feed themselves in a few hours? Or, to take a second example, Are the Lamellirostres, in spite of their desmognathous palates, more nearly allied to the schizognathous Gallo-Grallæ, because they agree with them in having the young born able to run and feed themselves in a few hours, than they are to the desmognathous Pelecano-Herodiones, which differ from them in having the young born helpless and requiring to be fed by their parents for many days?

Probably most ornithologists will, *à priori*, regard the condition of the young at birth as a less reliable character than the arrangement of the palatine bones and the feather-tracts; but if the relative importance of these characters is to be decided upon by the results of their use in the diagnoses of the subclasses into which birds may be divided, I for one am disposed to vote for the chicks and adopt the alternative scheme, which is elaborated on the next page.

Suborders.	Orders.	Subclasses.
1. Passeres.	PICO-PASSERES.	PASSERIFORMES.
2. Eurylæmi.		
3. Trochili.		
4. Scansores.		
5. Upupæ.		
6. Trogones.		
7. Columbæ.	COLUMBÆ.	
8. Musophagi.	COCCYGES.	
9. Cuculi.		
10. Halcyones.	PICARIÆ.	CORACIIFORMES.
11. Coraciæ.		
12. Bucerotes.		
13. Cathartes.	CATHARTES.	
14. Psittaci.	PSITTACI.	CICONIIFORMES.
15. Striges.	RAPTORES.	
16. Accipitres.		
17. Serpentarii.		
18. Plataleæ.	PELECANO-HERODIONES.	
19. Herodiones.		
20. Steganopodes		
21. Tubinares.	TUBINARES.	
22. Impennes.	IMPENNES.	
23. Palamedeæ.	LAMELLIROSTRES.	GALLIFORMES.
24. Anseres.		
25. Phœnicopter		
26. Gaviæ.	GALLO-GRALLÆ.	
27. Limicolæ.		
28. Grallæ.		
29. Fulicariæ.		
30. Pygopodes.		
31. Gallinæ.		
32. Crypturi.		
33. Apteryges.	APTERYGES.	STRUTHIONIFORMES.
34. Rheæ.	RATITÆ.	
35. Casuarii.		
36. Struthiones		

Subclass PASSERIFORMES.

So far as is known, the Passeriformes are the only birds which combine the following characters :—

A. Young born helpless, and requiring to be fed in the nest for many days, but not passing through a downy stage. **B**. A spinal feather-tract on a considerable part of the neck, well defined by lateral bare tracts and not split by a spinal bare tract. **C**. A hind toe is always present (generally the first digit, but occasionally, in a few Scansores which have lost the hallux, the fourth digit) and is connected with the *flexor longus hallucis*.

	1. Passeres.	2. Eurylæmi.	3. Trochili.	4. Scansores.	5. Upupæ.	6. Trogones.	7. Columbæ.	8. Musophagi.	9. Cuculi.	10. Serpentarii.	11. Accipitres.	12. Striges.	13. Psittaci.	14. Halcyones.	15. Coraciæ.	16. Bucerotes.	17. Cathartes.	18. Steganopodes.
A										*	*	*	*	…	…	…	*	*
B										…	…	…	…	…	…	*	*	*
C										…	…	…	…	*	*	*	*	…

	19. Herodiones.	20. Platalea.	21. Phœnicopteri.	22. Anseres.	23. Palamedeæ.	24. Tubinares.	25. Impennes.	26. Gaviæ.	27. Limicolæ.	28. Grallæ.	29. Fulicariæ.	30. Pygopodes.	31. Gallinæ.	32. Crypturi.	33. Apteryges.	34. Rheæ.	35. Casuarii.	36. Struthiones.
A	*	*	*	*	*	*	*	*	*	*	*	*	*	*	*	*	*	*
B	*	*	*	*	*	…	*	…	…	…	…	*	…	…	*	*	*	*
C	…	…	…	…	…	*	…	…	…	…	…	…	…	…	…	*	*	*

The condition of the young at birth, and their subsequent history, is a character of very great value in the classification of birds. Of this there can be no doubt whatever. However trivial the character may *à priori* appear, it is found upon investigation to be one of the least variable characters known.

The modifications in the condition of the young are as follows :—

Passerine.—Young born helpless and nearly naked; not passing through a downy stage before acquiring feathers.

Ardeine.—Young born helpless and nearly naked, but passing through a downy stage before acquiring feathers.

Accipitrine.—Young born helpless, but covered with down.

Galline.—Young born able to run, and covered with down.

Megapodiine.—Young born able to run, and covered with feathers.

The taxonomic value of these facts has been unfairly depreciated, in consequence of the exaggerated importance attached to some of them by Sundevall.

Order PICO-PASSERES.

The Pico-Passeres combine three characters not known to be combined in any other birds :—

A. Young born helpless and nearly naked; not passing through a downy stage before acquiring feathers. **B**. *Flexor longus hallucis* leading to one, or both, if two, hind toes (which is generally the hallux only, but is sometimes the fourth digit only, sometimes the first and fourth, and occasionally the first and second). **C**. Ambiens and accessory femoro-caudal muscles absent.

To this diagnosis other characters may be added :—

D. Spinal feather-tract well defined on the neck. **E**. Cervical vertebræ not exceeding 15 in number.

	1. Passeres.	2. Eurylæmi.	3. Trochili.	4. Scansores.	5. Upupæ.	6. Trogones.	7. Columbæ.	8. Musophagi.	9. Cuculi.	10. Serpentarii.	11. Accipitres.	12. Striges.	13. Psittaci.	14. Halcyones.	15. Coraciæ.	16. Bucerotes.	17. Cathartes.	18. Steganopodes.
A							...	...	...	*	*	*	*	...	...	...	*	*
B							...	...	*	...	...	*	*	*	*	*	*	...
C							*	*	*	*	*	...	...	...	...	...	*	...
D							...	...	...	...	...	...	...	...	...	*	*	*
E							...	...	...	...	...	...	...	...	...	...	...	...

	19. Herodiones.	20. Plataleæ.	21. Phœnicopteri.	22. Anseres.	23. Palamedeæ.	24. Tubinares.	25. Impennes.	26. Gaviæ.	27. Limicolæ.	28. Grallæ.	29. Fulicariæ.	30. Pygopodes.	31. Gallinæ.	32. Crypturi.	33. Apteryges.	34. Rheæ.	35. Casuarii.	36. Struthiones.
A	*	*	*	*	*	*	*	*	*	*	*	*	*	*	*	*	*	*
B	...	...	...	...	...	*	...	...	...	...	...	...	...	...	...	*	*	*
C	...	*	*	*	*	...	*	*	*	*	*	*	*	*	*	*	*	*
D	*	*	*	*	*	...	*	...	...	...	...	*	...	...	*	*	*	*
E	*	*	*	*	*	...	...	...	...	...	...	...	*	*	*	*	*	*

The number of vertebræ, in front of the first vertebra which is connected by a rib to the sternum, is not a very constant character, but is occasionally useful in diagnosis. In suborders 1 to 16 (with the doubtful exception of *Gyps bengalensis*) the number varies from 13 to 15. In suborders 19 to 23 and 32 to 36 the number varies from 16 to 21, except that in the Cygnidæ it reaches the extraordinary number of 23 to 25. Outside the Anseres the greatest variation appears to be in the Pygopodes—the Colymbidæ varying from 14 to 15, and the Podicipidæ from 17 to 21.

Suborder 1. PASSERES.

The Passeres may be diagnosed by two characters:—

A. Palate ægithognathous. **B**. Deep plantar tendons free.

To this diagnosis many other characters may be added:—

C. Oil-gland present, but nude. **D**. Spinal feather-tract well defined on neck. **E**. Spinal feather-tract not forked on the upper back. **F**. Hallux always present, and connected with the *flexor longus hallucis* and not with the *flexor perforans digitorum*.

	1. Passeres.	2. Eurylæmi.	3. Trochili.	4. Scansores.	5. Upupæ.	6. Trogones.	7. Columbæ.	8. Musophagi.	9. Cuculi.	10. Serpentarii.	11. Accipitres.	12. Striges.	13. Psittaci.	14. Halcyones.	15. Coraciæ.	16. Bucerotes.	17. Cathartes.	18. Steganopodes.
A		...	*	...	*	*	*	*	*	*	*	*	*	*	...	*	*	*
B		*	*	*	...	*	*	*	*	*	*	*	*	*	*	*	*	*
C		...	...	...	*	...	...	*	...	*	*	...	*	...	...	*	...	*
D		...	...	...	...	...	...	...	...	...	...	...	...	...	...	*	*	*
E		...	...	...	*	...	*	...	*	*	*	*	*	...	*	...	*	...
F		...	...	...	...	...	...	...	...	...	...	...	...	*	*	*	*	...

	19. Herodiones.	20. Plataleæ.	21. Phœnicopteri.	22. Anseres.	23. Palamedeæ.	24. Tubinares.	25. Impennes.	26. Gaviæ.	27. Limicolæ.	28. Grallæ.	29. Fulicariæ.	30. Pygopodes.	31. Gallinæ.	32. Crypturi.	33. Apteryges.	34. Rheæ.	35. Casuarii.	36. Struthiones.
A	*	*	*	*	*	*	*	...	...	...	*	*	*	*	*	*	*	*
B	...	*	*	*	*	*	*	*	*	*	*	*	*	*	*	*	*	*
C	*	*	*	*	*	*	*	*	*	...	...	*	...	*	*	*	*	*
D	*	*	*	*	*	...	*	...	...	...	...	*	...	...	*	*	*	*
E	...	...	...	...	...	*	...	*	*	*	*	...	...	...	...	...	...	...
F	...	...	...	...	...	*	...	...	...	...	...	...	...	...	...	*	*	*

The Passeres may be divided into at least five families, which may be diagnosed by satisfactory characters (Garrod, Proc. Zool. Soc. 1877, p. 518). It is impossible to say to what extent the subfamilies are capable of being diagnosed.

The pterylosis of the hind neck appears to be a character of considerable taxonomic value.

The spinal feather-tract may be regarded as well defined on the neck when the lateral bare tracts extend for some distance up each side of the neck, and when it is not split and consequently displaced by a spinal bare tract.

The spinal feather-tract may be regarded as not defined on the neck either (*a*) when it coalesces with the ventral feather-tracts in the absence of lateral bare tracts, or (*b*) when a spinal bare tract extends up the neck as far as or further than the lateral bare tracts.

As thus defined these two types are very constant, the only exceptions known being in the Limicolæ, Grallæ, and Fulicariæ.

Suborder 2. EURYLÆMI.

The Broadbills may be diagnosed as follows :—

A. Palate ægithognathous. **B**. Young born helpless and not passing through a downy stage. **C**. *Flexor longus hallucis* leading to hallux after sending down a tendon to the *flexor perforans digitorum*, which leads to three front digits.

To this diagnosis may be added many other characters :—

D. Oil-gland present, but nude. **E**. Nasals holorhinal. **F**. Dorsal vertebræ heterocœlous.

Characters **D** and **E**, or **E** and **F**, may be substituted for character **B** if preferred.

	1. Passeres.	2. Eurylæmi.	3. Trochili.	4. Scansores.	5. Upupæ.	6. Trogones.	7. Columbæ.	8. Musophagi.	9. Cuculi.	10. Serpentarii.	11. Accipitres.	12. Striges.	13. Psittaci.	14. Halcyones.	15. Coraciæ.	16. Bucerotes.	17. Cathartes.	18. Steganopodes.	19. Herodiones.	20. Plataleæ.	21. Phœnicopteri.	22. Anseres.	23. Palamedeæ.	24. Tubinares.	25. Impennes.	26. Gaviæ.	27. Limicolæ.	28. Grallæ.	29. Fulicariæ.	30. Pygopodes.	31. Gallinæ.	32. Crypturi.	33. Apteryges.	34. Rheæ.	35. Casuarii.	36. Struthiones.
A	...		*	...	*	*	*	*	*	*	*	*	*	*	...	*	*	*	*	*	*	*	*	*	*	...	...	...	*	*	*	*	*	*	*	*
B	...		...	...	...	...	...	...	...	*	*	*	*	...	...	...	*	*	*	*	*	*	*	*	*	*	*	*	*	*	*	*	*	*	*	*
C	*		...	*	*	*	...	...	...	...	...	...	...	*	*	*	*	...	...	...	...	...	...	*	...	...	...	...	...	...	...	...	...	*	*	*
D	...		...	...	*	...	...	*	...	*	*	...	*	...	...	*	*	*	*	*	*	*	*	*	*	*	*	...	...	*	...	*	*	*	*	*

The arrangement of the deep plantar tendons appears to me to have much more taxonomic value than has been attached to it. The modifications to which it is subject may be classified as follows :—

Passerine.—Front plantar (*flexor perforans digitorum*) splitting into three tendons at the foot of the tarsus to lead to the three front toes (second, third, and fourth digits). Hind plantar (*flexor longus hallucis*) leading to hallux only, and not connected with front plantar, either by a fibrous vinculum or a tendon.

Galline.—Front plantar as in the Passerine arrangement. Hind plantar connected with front plantar either by a fibrous vinculum or by a tendon, and leading either (*a*) to hallux only, (*b*) to first and second digits, (*c*) to first, second, and third digits, (*d*) to all four digits, or, where the hallux is absent or very small, (*e*) coalescing with the front plantar.

Trogonine.—Front plantar leading to the two front toes (third and fourth digits). Hind plantar leading to the two hind toes (first and second digits), and also sending down a tendon to front plantar.

Picine.—Front plantar leading to third digit only. Hind plantar leading to first, second, and fourth digits, and also sending down a tendon to front plantar.

Coraciine.—Plantars coalesced, but front plantar (sometimes before, sometimes after coalescence) leading to hallux. Front plantar leading either (*a*) to first, second, and third digits, or (*b*) to all four digits. Hind plantar leading either (*a*) to third and fourth digits, or (*b*) to second, third, and fourth digits.

Suborder 3. TROCHILI.

The Humming-Birds may be diagnosed as follows :—

A. Young born helpless and not passing through a downy stage. **B.** Palate schizognathous. **C.** Basipterygoid processes absent. **D.** Nasals holorhinal. **E.** Front plantar leading to three front toes; hind plantar leading to hallux.

This diagnosis is complete without character **C**.

	1. Passeres.	2. Eurylæmi.	3. Trochili.	4. Scansores.	5. Upupæ.	6. Trogones.	7. Columbæ.	8. Musophagi.	9. Cuculi.	10. Serpentarii.	11. Accipitres.	12. Striges.	13. Psittaci.	14. Halcyones.	15. Coraciæ.	16. Bucerotes.	17. Cathartes.	18. Steganopodes.	19. Herodiones.	20. Plataleæ.	21. Phœnicopteri.	22. Anseres.	23. Palamedeæ.	24. Tubinares.	25. Impennes.	26. Gaviæ.	27. Limicolæ.	28. Grallæ.	29. Fulicariæ.	30. Pygopodes.	31. Gallinæ.	32. Crypturi.	33. Apteryges.	34. Rheæ.	35. Casuarii.	36. Struthiones.
A	...	...		...	...	...	...	...	...	*	*	*	*	...	...	...	*	*	*	*	*	*	*	*	*	*	*	*	*	*	*	*	*	*	*	*
B	*	*		...	*	...	...	*	*	*	*	*	*	*	...	*	*	*	*	*	*	*	*	...	...	...	...	...	...	...	...	*	*	*	*	*
C	...	...		...	...	*	*	...	...	*	...	*	...	...	...	...	*	...	...	...	...	*	*	...	...	...	*	...	...	...	*	*	*	*	*	*
D	...	...		...	...	...	*	...	...	...	...	...	...	...	...	...	...	...	...	*	...	...	...	...	...	...	*	*	...	...	...	*	...	...	...	...
E	...	...		*	...	*	...	...	...	...	...	...	...	*	*	*	*	...	...	...	...	...	...	*	...	...	...	...	...	...	...	...	...	*	*	*

The Humming-Birds, the Swifts, the Owls, and the Pelicans are the only birds known to have neither ambiens, accessory femoro-caudal, semitendinosus, nor accessory semitendinosus muscles; but it is not to be supposed that this coincidence denotes relationship in any way.

Garrod and Forbes attached great importance to the bifurcation of the nasal bone, of which they recognized two modifications :—

Holorhinal.—Angle of bifurcation rounded off and placed forwards, so that a line drawn across the skull at a tangent to the two curves falls in front of the lachrymals and the termination of the nasal processes of the premaxillary.

Schizorhinal.—Angle of bifurcation in most cases very acute, but occasionally rounded; always placed backwards, so that the apex extends as far back as the posterior terminations of the nasal processes of the premaxillary between the centres of the lachrymals.

These characters are only known to break down in the Passeres and the Gaviæ.

Suborder 4. SCANSORES.

The Woodpeckers and their allies may be diagnosed in various ways. The following character is diagnostic:—

A. *Flexor perforans digitorum* leading to third digit only; *flexor longus hallucis* first sending a tendon to the other plantar, then a second to fourth digit, after which (if hallux present) splitting into two tendons, one leading to hallux, the other to second digit.

The combination of the following two characters is also diagnostic:—

B. Feet zygodactyle. **C.** Spinal feather-tract well defined on the neck, and forked on the lower (not on the upper) back.

To these characters may be added:—

D. Basipterygoid processes absent.

	1. Passeres.	2. Eurylæmi.	3. Trochili.	4. Scansores.	5. Upupæ.	6. Trogones.	7. Columbæ.	8. Musophagi.	9. Cuculi.	10. Serpentarii.	11. Accipitres.	12. Striges.	13. Psittaci.	14. Halcyones.	15. Coraciæ.	16. Bucerotes.	17. Cathartes.	18. Steganopodes.
A	*	*	*		*	*	*	*	*	*	*	*	*	*	*	*	*	*
B	*	*	*		*	*	*	...	...	*	...	...	...	*	...	*	*	*
C	...	...	...		*	*	*	*	*	*	*	*	*	...	*	*	*	*
D	...	...	...		...	*	*	...	...	*	...	*	...	...	...	...	*	...

	19. Herodiones.	20. Plataleæ.	21. Phœnicopteri.	22. Anseres.	23. Palamedeæ.	24. Tubinares.	25. Impe ines.	26. Gaviæ.	27. Limicolæ.	28. Grallæ.	29. Fulicariæ.	30. Pygopodes.	31. Gallinæ.	32. Crypturi.	33. Apteryges.	34. Rheæ.	35. Casuarii.	36. Stru.hiones.
A	*	*	*	*	*	*	*	*	*	*	*	*	*	*	*	*	*	*
B	*	*	*	*	*	*	*	*	*	*	*	*	*	*	*	*	*	*
C	*	*	*	*	*	...	*	*	*	*	*	*	...	*	*	*	*	*
D	...	...	...	*	*	...	...	...	*	...	...	...	*	*	*	*	*	*

The Scansores consist of six families, which differ from each other as follows. They may be divided into two groups:—

a. Passerine Scansores: *with cœca and a nude oil-gland.*

Galbulidæ.—No vomer; a narrow clavicular feather-tract on each side of breast; a small aftershaft.

Bucconidæ.—Vomer present; no clavicular feather-tract; no aftershaft.

b. Picine Scansores: *with no cœca, but with a tufted oil-gland.*

Picidæ.—Palate schizognathous; vomer slender, pointed, split.

Indicatoridæ.—Palate ægithognathous; ventral feather-tract forked on throat, but not on each side of breast; vomer bifid. (Garrod, Proc. Zool. Soc. 1878, p. 930.)

Capitonidæ.—Palate sometimes ægithognathous, sometimes desmognathous; vomer bifid; ventral feather-tract forked on throat and on each side of breast.

Rhamphastidæ.—Palate desmognathous; vomer truncated; postscapular feather-tracts separated on lower back and rump. (Garrod, Proc. Zool. Soc. 1878, p. 462.)

Suborder 5. UPUPÆ.

The Hoopoes and their allies may be diagnosed as follows:—

A. Deep plantar tendons free. **B**. Episternal process perforated to receive the feet of the coracoids.

To this diagnosis may be added many other characters:—

C. Palate desmognathous. **D**. Oil-gland tufted. **E**. Spinal feather-tract forked on the upper back.

	1. Passeres.	2. Eurylæmi.	3. Trochili.	4. Scansores.	5. Upupæ.	6. Trogones.	7. Columbæ.	8. Musophagi.	9. Cuculi.	10. Serpentarii.	11. Accipitres.	12. Striges.	13. Psittaci.	14. Halcyones.	15. Coraciæ.	16. Bucerotes.	17. Cathartes.	18. Steganopodes.	19. Herodiones.	20. Plataleæ.	21. Phœnicopteri.	22. Anseres.	23. Palamedeæ.	24. Tubinares.	25. Impennes.	26. Gaviæ.	27. Limicolæ.	28. Grallæ.	29. Fulicariæ.	30. Pygopodes.	31. Gallinæ.	32. Crypturi.	33. Apteryges.	34. Rheæ.	35. Casuarii.	36. Struthiones.
A	...	*	*	*		*	*	*	*	*	*	*	*	*	*	*	*	*	...	*	*	*	*	*	*	*	*	*	*	*	*	*	*	*	*	*
B	*	*	*	*		*	*	*	*	*	*	*	*	...	*	..	*	*	*	*	*	*	*	*	*	*	*	*	*	*	...	*	*	*	*	*
C	*	*	*	...		*	*	...	...	...	...	...	...	...	...	...	...	...	...	...	...	...	...	*	*	*	*	*	*	*	*	*	*	*	*	*
D	*	*	*	...		*	*	...	*	...	...	*	...	...	*	...	*	...	...	...	...	...	...	...	...	...	...	...	...	...	...	...	*	*	*	*
E	*	*	*	*		*	...	*	...	...	...	...	...	*	...	...	...	...	...	...	...	...	...	...	*	...	...	...	...	...	...	*	*	*	*	*

The Upupæ may probably be divided into two families—the Upupidæ and the Irrisoridæ.

We have already had occasion to note the importance of the existence of a spinal feather-tract on the neck. A second character of importance to be found in the pterylosis of the upper parts is, that in some birds the spinal feather-tract continues uninterruptedly down the neck, and is not split by a spinal bare tract on the upper back; whilst in others it is so split. Amongst the birds which have a well-defined spinal feather-tract we may recognize three modifications:—

Accipitrine.—Spinal feather-tract forked on the upper back.

Hirundine.—Spinal feather-tract forked on the lower back.

Sylviine.—Spinal feather-tract not forked at all, though widening into a saddle on the lower back.

Suborder 6. TROGONES.

The Trogons have the second digit reversed. The front plantar splits into two to lead to the two front toes, and the hind plantar does the same to lead to the two hind toes. Either of these characters is diagnostic. The combination of the following characters is also diagnostic:—

A. Palate schizognathous. **B.** Basipterygoid processes present. **C.** Spinal feather-tract well defined from nape to oil-gland, and not forked. **D.** Young born helpless and not passing through a downy stage.

	1. Passeres.	2. Eurylæmi.	3. Trochili.	4. Scansores.	5. Upupæ.	6. Trogones.	7. Columbæ.	8. Musophagi.	9. Cuculi.	10. Serpentarii.	11. Accipitres.	12. Striges.	13. Psittaci.	14. Halcyones.	15. Coraciæ.	16. Bucerotes.	17. Cathartes.	18. Steganopodes.
A	*	*	...	...	*		...	*	*	*	*	*	*	*	...	*	*	*
B	*	*	*	*	*		...	*	*	...	*	...	*	*	...	*	...	*
C	...	*	*	*	*		*	...	*	*	*	*	*	...	*	...	*	*
D	...	...	...	...	...		...	...	...	*	*	*	*	...	...	...	*	*

	19. Herodiones.	20. Plataleæ.	21. Phœnicopteri.	22. Anseres.	23. Palamedeæ.	24. Tubinares.	25. Impennes.	26. Gaviæ.	27. Limicolæ.	28. Grallæ.	29. Fulicariæ.	30. Pygopodes.	31. Gallinæ.	32. Crypturi.	33. Apteryges.	34. Rheæ.	35. Casuarii.	36. Struthiones.
A	*	*	*	*	*	...	...	...	...	...	...	...	...	*	*	*	*	*
B	*	*	*	...	...	...	*	*	...	...	...	*	...	...	...	...	...	...
C	*	*	*	*	*	*	*	*	*	*	*	*	...	...	*	*	*	*
D	*	*	*	*	*	*	*	*	*	*	*	*	*	*	*	*	*	*

This is the first appearance of basipterygoid processes, which we may assume to have once been universal, inasmuch as they are found in the embryos of many Passeres.

The presence or absence of basipterygoid processes on the basisphenoid, or upon its rostrum, which articulate with facets on the pterygoids, is a character of great value in the diagnosis of many groups of birds; but the following suborders contain species in which basipterygoid processes are present, as well as others in which they are absent:—Coraciæ, Tubinares, Grallæ, and Fulicariæ.

Order **COLUMBÆ.**

Now that the Dididæ have become extinct, the order Columbæ contains only one suborder.

Suborder 7. COLUMBÆ.

The Columbæ combine three characters that are not known to be combined in any other birds :—

A. Young born helpless and not passing through a downy stage. **B**. Palate schizognathous. **C**. Nasals schizorhinal.

To these may be added many others :—

D. Basipterygoid processes present.

	1. Passeres.	2. Eurylæmi.	3. Trochili.	4. Scansores.	5. Upupæ.	6. Trogones.	7. Columbæ.	8. Musophagi.	9. Cuculi.	10. Serpentarii.	11. Accipitres.	12. Striges.	13. Psittaci.	14. Halcyones.	15. Coraciæ.	16. Bucerotes.	17. Cathartes.	18. Steganopodes.	19. Herodiones.	20. Plataleæ.	21. Phœnicopteri.	22. Anseres.	23. Palamedeæ.	24. Tubinares.	25. Impennes.	26. Gaviæ.	27. Limicolæ.	28. Grallæ.	29. Fulicariæ.	30. Pygopodes.	31. Gallinæ.	32. Crypturi.	33. Apteryges.	34. Rheæ.	35. Casuarii.	36. Struthiones.
A	...	...	...	...	...	...		...	...	*	*	*	*	...	...	...	*	*	*	*	*	*	*	*	*	*	*	*	*	*	*	*	*	*	*	*
B	*	*	...	...	*	...		*	*	*	*	*	*	*	...	*	*	*	*	*	*	*	*	...	...	...	...	...	...	...	...	*	*	*	*	*
C	...	*	*	*	*	*		*	*	*	*	*	*	*	*	*	*	*	*	...	*	*	*	*	*	...	...	...	*	*	*	...	*	*	*	*
D	*	*	*	*	*	...		*	*	...	*	...	*	*	...	*	...	*	*	*	*	...	...	...	*	*	...	...	...	*	...	...	...	...	...	...

The ambiens and accessory femoro-caudal muscles make their first appearance in the Columbæ; but in some Pigeons the former is absent, and in a few the latter, though it is not known that in any species both are absent.

The presence or absence of the ambiens muscle, a character to which Garrod and Forbes attached primary value, is not to be rejected as worthless, though it breaks down in the following suborders, each of which contain some species with an ambiens muscle and others without:—Columbæ, Psittaci, Steganopodes, Herodiones, Tubinares, Gaviæ, Pygopodes, and Casuarii.

The presence or absence of the accessory femoro-caudal muscle is apparently of more taxonomic value, as it is only known to break down in the following suborders:—Columbæ, Cuculi, Tubinares, Limicolæ, and Grallæ.

Order COCCYGES.

The Coccyges may be diagnosed as follows:—

A. Palate desmognathous. **B.** Basipterygoid processes absent. **C.** Hallux always present, and connected with the *flexor longus hallucis*, and not with the *flexor perforans digitorum*, which leads to the second, third, and fourth digits. **D.** Young born helpless and not passing through a downy stage before acquiring feathers.

	1. Passeres.	2. Eurylæmi.	3. Trochili.	4. Scansores.	5. Upupæ.	6. Trogones.	7. Columbæ.	8. Musophagi.	9. Cuculi.	10. Serpentarii.	11. Accipitres.	12. Striges.	13. Psittaci.	14. Halcyones.	15. Coraciæ.	16. Bucerotes.	17. Cathartes.	18. Steganopodes.	19. Herodiones.	20. Plataleæ.	21. Phœnicopteri.	22. Anseres.	23. Palamedeæ.	24. Tubinares.	25. Impennes.	26. Gaviæ.	27. Limicolæ.	28. Grallæ.	29. Fulicariæ.	30. Pygopodes.	31. Gallinæ.	32. Crypturi.	33. Apteryges.	34. Rheæ.	35. Casuarii.	36. Struthiones.
A	*	*	*	...	...	*	*			...	...	...	...	...	...	...	...	...	...	...	...	...	...	*	*	*	*	*	*	*	*	*	*	*	*	*
B	...	...	...	...	...	*	*			*	...	*	...	...	..	...	*	...	...	...	...	*	*	...	...	...	*	...	..	...	*	*	*	*	*	*
C	*	...	...	*	*	*	...			...	...	...	...	*	*	*	*	...	...	...	...	...	...	*	...	...	...	...	...	...	...	...	..	*	*	*
D	...	...	...	...	...	...	...			*	*	*	*	...	...	...	*	*	*	*	*	*	*	*	*	*	*	*	*	*	*	*	*	*	*	*

The arrangement of the bones of the palate, with which Huxley's name is especially associated, is of great value as a taxonomic character, though it breaks down occasionally, like every other character. The following suborders contain species belonging to more than one of the modifications which are regarded as important:—Scansores, Coraciæ, Gaviæ, and Grallæ. These modifications may be described as follows:—

Ægithognathous.—Maxillo-palatines not coalesced with each other across the middle line; vomer broad and truncated in front.

Schizognathous.—Maxillo-palatines not coalesced with each other across the middle line; vomer narrow and more or less pointed in front.

Desmognathous.—Maxillo-palatines either (*a*) coalesced with each other across the middle line, or (*b*) united by an ossified nasal septum.

Dromæognathous.—Vomer expanded behind so as to prevent the palatines from articulating with the rostrum of the basisphenoid.

Suborder 8. MUSOPHAGI.

The Plantain-Eaters may be diagnosed as follows :—

A. Palate desmognathous. **B**. Plantars galline. **C**. Spinal feather-tract well defined on neck, and not forked on back.

Other characters may be added :—

D. Oil-gland tufted.

	1. Passeres.	2. Eurylæmi.	3. Trochili.	4. Scansores.	5. Upupæ.	6. Trogones.	7. Columbæ.	8. Musophagi.	9. Cuculi.	10. Serpentarii.	11. Accipitres.	12. Striges.	13. Psittaci.	14. Halcyones.	15. Coraciæ.	16. Bucerotes.	17. Cathartes.	18. Steganopodes.
A	*	*	*	...	...	*	*		...	...	...	...	...	...	...	...	...	...
B	*	...	...	*	*	*	...		...	...	...	...	...	*	*	*	*	...
C	...	*	*	*	*	...	*		*	*	*	*	*	...	*	...	*	*
D	*	*	*	...	...	*	*		*	...	...	*	...	...	*	...	*	...

	19. Herodiones.	20. Plataleæ.	21. Phœnicopteri.	22. Anseres.	23. Palamedeæ.	24. Tubinares.	25. Impennes.	26. Gaviæ.	27. Limicolæ.	28. Grallæ.	29. Fulicariæ.	30. Pygopodes.	31. Gallinæ.	32. Crypturi.	33. Apteryges.	34. Rheæ.	35. Casuarii.	36. Struthiones.
A	...	...	...	...	...	*	*	*	*	*	*	*	*	*	*	*	*	*
B	...	...	...	...	...	...	...	...	...	...	...	...	...	...	...	...	...	...
C	*	*	*	*	*	*	*	*	*	*	*	*	...	...	*	*	*	*
D	...	...	...	...	...	...	...	...	...	...	...	...	...	...	*	*	*	*

The oil-gland may be (*a*) absent or present; and when present may be either (*b*) surrounded by a tuft of feathers, or (*c*) nude. This character, to which Garrod and Forbes attached great importance, is fairly constant, but breaks down in the following suborders:— Scansores, Columbæ, Psittaci, Halcyones, Coraciæ, Grallæ, Fulicariæ, and Gallinæ.

Suborder 9. CUCULI.

The Cuckoos may be diagnosed as follows :—

A. Basipterygoid processes absent. **B.** Feet zygodactyle. **C.** Plantars galline. **D.** Oil-gland present, but nude.

Other characters may be added :—

E. Palate desmognathous.

	1. Passeres.	2. Eurylæmi.	3. Trochili.	4. Scansores.	5. Upupæ.	6. Trogones.	7. Columbæ.	8. Musophagi.	9. Cuculi.	10. Serpentarii.	11. Accipitres.	12. Striges.	13. Psittaci.	14. Halcyones.	15. Coraciæ.	16. Bucerotes.	17. Cathartes.	18. Steganopodes.	19. Herodiones.	20. Plataleæ.	21. Phœnicopteri.	22. Anseres.	23. Palamedeæ.	24. Tubinares.	25. Impennes.	26. Gaviæ.	27. Limicolæ.	28. Grallæ.	29. Fulicariæ.	30. Pygopodes.	31. Gallinæ.	32. Crypturi.	33. Apteryges.	34. Rheæ.	35. Casuarii.	36. Struthiones.
A	...	...	...	...	...	*	*	...		*	...	*	...	...	...	...	*	...	...	...	...	*	*	...	...	...	*	...	...	...	*	*	*	*	*	*
B	*	*	*	...	*	*	*	...		*	...	...	...	*	...	*	*	*	*	*	*	*	*	*	*	*	*	*	*	*	*	*	*	*	*	*
C	*	...	...	*	*	*	...	...		...	...	...	...	*	*	*	*	...	...	...	...	...	...	...	...	...	...	...	...	...	...	...	...	...	...	...
D	...	...	...	...	*	...	...	*		*	*	...	*	...	...	*	*	*	*	*	*	*	*	*	*	*	*	...	...	*	...	*	*	*	*	*
E	*	*	*	...	...	*	*	...		...	...	...	...	...	...	...	...	...	...	...	...	...	...	*	*	*	*	*	*	*	*	*	*	*	*	*

The Cuculi have zygodactyle feet, but they are correlated with galline plantars.

In zygodactyle feet the fourth digit is turned backwards as well as the hallux. The first digit is, however, occasionally absent. The following suborders contain species which may be regarded as semi-zygodactyle, inasmuch as the fourth digit may be turned either backwards or forwards at pleasure :—Musophagi, Accipitres, Striges, and Coraciæ.

Subclass FALCONIFORMES.

The Falconiformes are supposed to be the only birds which combine the following characters :—

A. Young born helpless, but passing through a downy stage before acquiring feathers. **B.** Hallux always present, and connected with the *flexor longus hallucis*, and not with the *flexor perforans digitorum*. **C.** Spinal feather-tract well defined on the neck.

	1. Passeres.	2. Eurylæmi.	3. Trochili.	4. Scansores.	5. Upupæ.	6. Trogones.	7. Columbæ.	8. Musophagi.	9. Cuculi.	10. Serpentarii.	11. Accipitres.	12. Striges.	13. Psittaci.	14. Halcyones.	15. Coraciæ.	16. Bucerotes.	17. Cathartes.	18. Steganopodes.	19. Herodiones.	20. Plataleæ.	21. Phœnicopteri.	22. Anseres.	23. Palamedeæ.	24. Tubinares.	25. Impennes.	26. Gaviæ.	27. Limicolæ.	28. Grallæ.	29. Fulicariæ.	30. Pygopodes.	31. Gallinæ.	32. Crypturi.	33. Apteryges.	34. Rheæ.	35. Casuarii.	36. Struthiones.
A	*	*	*	*	*	*	*	*	*					*	...	*	...	...	...	...	*	*	*	...	...	*	*	*	*	*	*	*	*	*	*	*
B	...	...	...	...	...	...	...	...	...					*	*	*	*	...	...	...	...	...	...	*	...	...	...	...	...	...	...	...	...	*	*	*
C	...	...	...	...	...	...	...	...	...					...	...	*	*	*	*	*	*	*	*	...	*	...	...	...	...	*	...	...	*	*	*	*

The Falconiformes are allied to the Passeriformes, and do not differ from them in any very important feature, except that the young birds pass through a downy stage before acquiring feathers. It is possible that this character may be of older date, and consequently of greater taxonomic value, than the definition of the feather-tracts on the neck or the ossification of the nasal septum—in which case the Falconiformes might be added to the Pelecano-Herodiones, the Tubinares, and the Impennes. An alternative arrangement of the various Orders and Suborders based upon this hypothesis will be found in an appendix to the Introduction.

Order **RAPTORES.**

The Raptores possess three characters which are not known to be combined in any other birds:—

A. Young born helpless, but covered with down. **B.** Hallux always present, and connected with the *flexor longus hallucis*, and not with the *flexor perforans digitorum*. **C.** Spinal feather-tract well defined on the neck.

	1. Passeres.	2. Eurylæmi.	3. Trochili.	4. Scansores.	5. Upupæ.	6. Trogones.	7. Columbæ.	8. Musophagi.	9. Cuculi.	10. Serpentarii.	11. Accipitres.	12. Striges.	13. Psittaci.	14. Halcyones.	15. Coraciæ.	16. Bucerotes.	17. Cathartes.	18. Steganopodes.	19. Herodiones.	20. Plataleæ.	21. Phœnicopteri.	22. Anseres.	23. Palamedeæ.	24. Tubinares.	25. Impennes.	26. Gaviæ.	27. Limicolæ.	28. Grallæ.	29. Fulicariæ.	30. Pygopodes.	31. Gallinæ.	32. Crypturi.	33. Apteryges.	34. Rheæ.	35. Casuarii.	36. Struthiones.
A	*	*	*	*	*	*	*	*	*				*	*	...	*	...	*	...	...	*	*	*	...	...	*	*	*	*	*	*	*	*	*	*	*
B	...	...	...	...	...	...	...	...	...				...	*	*	*	*	...	...	...	...	...	...	*	...	...	...	...	...	...	...	...	...	*	*	*
C	...	...	...	...	...	...	...	...	...				...	...	...	*	*	*	*	*	*	*	*	...	*	...	...	...	...	*	...	...	*	*	*	*

The admission or rejection of the Striges into the Raptores has been a vexed question for some time. Fürbringer places the Striges in his Coraciiformes, next to the Podargidæ. In the following table some of the characters are compared, the relation of the Striges to the Podargidæ on the one hand and to the Accipitres on the other being expressed:—

Podargidæ.		*Accipitres.*
differ	maxillo-palatines	agree
do.	palatines	do.
do.	basipterygoids	do.
do.	plantars	do.
agree	ambiens	differ
do.	femoro-caudal	agree
do.	accessory do.	do.
differ	semitendinosus	do
do.	accessory do.	do.
agree	cæca	do.
differ	oil-gland	differ

Suborder 10. SERPENTARII.

The Secretary Bird may be diagnosed as follows :—

A. Palate desmognathous. **B.** Basipterygoid processes present. **C.** Spinal feather-tract well defined on the neck. **D.** Deep plantar tendons galline. **E.** Oil-gland tufted.

	1. Passeres.	2. Eurylæmi.	3. Trochili.	4. Scansores.	5. Upupæ.	6. Trogones.	7. Columbæ.	8. Musophagi.	9. Cuculi.	10. Serpentarii.	11. Accipitres.	12. Striges.	13. Psittaci.	14. Halcyones.	15. Coraciæ.	16. Bucerotes.	17. Cathartes.	18. Steganopodes.	19. Herodiones.	20. Plataleæ.	21. Phœnicopteri.	22. Anseres.	23. Palamedeæ.	24. Tubinares.	25. Impennes.	26. Gaviæ.	27. Limicolæ.	28. Grallæ.	29. Fulicariæ.	30. Pygopodes.	31. Gallinæ.	32. Crypturi.	33. Apteryges.	34. Rheæ.	35. Casuarii.	36. Struthiones.
A	*	*	*	...	...	*	*	...	...		...	...	...	...	...	...	...	...	...	...	...	...	...	*	*	*	*	*	*	*	*	*	*	*	*	*
B	*	*	*	*	*	...	...	*	*		*	...	*	*	...	*	...	*	*	*	*	...	...	...	*	*	...	...	...	*	...	...	...	...	...	...
C	...	...	...	...	...	...	...	...	...		...	...	...	...	...	*	*	*	*	*	*	*	*	...	*	...	...	...	...	*	...	...	*	*	*	*
D	*	...	...	*	*	*	...	...	...		...	...	...	*	*	*	*	...	...	...	...	...	...	...	...	...	...	...	...	...	...	...	...	...	...	...
E	*	*	*	...	...	*	*	...	*		...	*	...	...	*	...	*	...	...	...	...	...	...	...	...	...	...	...	...	...	...	...	*	*	*	*

The Secretary Bird further differs from the Accipitres in having no femoro-caudal muscle, in having an accessory femoro-caudal, and in having a semitendinosus and an accessory semitendinosus.

Some ornithologists regard *Secretarius* as very closely allied to *Cariama*, which I have placed in the Fulicariæ or Holorhinal Grallæ. That it should be possible for any difference of opinion to exist on the subject is a proof of the feebleness of the characters which separate the Orders of Birds. The Raptores have an ossified nasal septum which divides the nostrils, and the lateral bare tracts run far up the neck. In both cases *Cariama* agrees with the majority of the Fulicariæ, and differs from all the Raptores. Unfortunately the young of *Cariama* is unknown. If they are born able to run, *Cariama* belongs to the Gallo-Grallæ; if they are fed for some days in the nest, it ought possibly to be placed in an Order between the Raptores and the Psittaci; but probably the subclass Falconiformes could not then be diagnosed.

Suborder 11. ACCIPITRES.

The Birds of Prey, after the Owls (Striges), the American Vultures (Cathartes), and the Secretary Bird (Serpentarii) have been excluded, may be diagnosed as follows :—

A. Young born helpless, but covered with down. **B.** Basipterygoid processes absent. **C.** Spinal feather-tract well defined on the neck. **D.** Hallux always present, and connected with the *flexor longus hallucis*, and not with the *flexor perforans digitorum*; the two tendons bound together by a fibrous vinculum.

A simpler, though much less satisfactory, diagnosis is as follows :—

E. Myological formula A+. **F.** Dorsal vertebræ heterocœlous.

The first of these characters (which implies that the ambiens and femoro-caudal muscles are present, whilst the accessory femoro-caudal, semitendinosus, and accessory semitendinosus are absent) excludes all birds except the Fregatidæ (a family in the suborder Steganopodes), which are excluded by the second character.

	1. Passeres.	2. Eurylæmi.	3. Trochili.	4. Scansores.	5. Upupæ.	6. Trogones.	7. Columbæ.	8. Musophagi.	9. Cuculi.	10. Serpentarii.	11. Accipitres.	12. Striges.	13. Psittaci.	14. Halcyones.	15. Coraciæ.	16. Bucerotes.	17. Cathartes.	18. Steganopodes.	19. Herodiones.	20. Plataleæ.	21. Phœnicopteri.	22. Anseres.	23. Palamedeæ.	24. Tubinares.	25. Impennes.	26. Gaviæ.	27. Limicolæ.	28. Grallæ.	29. Fulicariæ.	30. Pygopodes.	31. Gallinæ.	32. Crypturi.	33. Apteryges.	34. Rheæ.	35. Casuarii.	36. Struthiones.
A	*	*	*	*	*	*	*	*	*	...		...	*	*	...	*	...	*	...	...	*	*	*	...	...	*	*	*	*	*	*	*	*	*	*	*
B	...	...	...	...	...	*	*	...	...	*		*	...	...	...	...	*	...	...	...	...	*	*	...	...	...	*	...	...	...	*	*	*	*	*	*
C	...	...	...	..	...	...	...	...	...	...		...	...	...	...	*	*	*	*	*	*	*	*	...	*	...	...	...	...	*	...	...	*	*	*	*
D	*	...	...	*	*	*	...	...	...	...		...	...	*	*	*	*	. .	...	...	...	...	...	*	...	...	...	...	..	...	...	...	...	*	*	*

The vertebræ of most birds are heterocœlous or saddle-shaped—that is to say, the facets, where they articulate with the adjoining vertebræ, are convex in front and concave behind when seen in vertical section, but are concave in front and convex behind when seen in lateral section. In a few birds, however, the dorsal vertebræ, especially the hinder ones, are more or less opisthocœlous—that is to say, the facets, where they articulate with the adjoining vertebræ, are either straight or concave (not convex) behind when seen in lateral section.

This character breaks down in the Coraciæ and the Steganopodes, in both of which sub-orders families with opisthocœlous dorsal vertebræ occur, together with others in which the dorsal vertebræ are heterocœlous.

Suborder 12. STRIGES.

The Owls may be diagnosed as follows:—

A. Young born helpless, but covered with down. **B.** Oil-gland present, but nude. **C.** Spinal feather-tract well defined on the neck.

To this diagnosis many other characters may be added:—

D. Basipterygoid processes present. **E.** *Flexor longus hallucis* leading to hallux, *flexor perforans digitorum* to second, third, and fourth digits.

	1. Passeres.	2. Eurylæmi.	3. Trochili.	4. Scansores.	5. Upupæ.	6. Trogones.	7. Columbæ.	8. Musophagi.	9. Cuculi.	10. Serpentarii.	11. Accipitres.	12. Striges.	13. Psittaci.	14. Halcyones.	15. Coraciæ.	16. Bucerotes.	17. Cathartes.	18. Steganopodes.	19. Herodiones.	20. Plataleæ.	21. Phœnicopteri.	22. Anseres.	23. Palamedeæ.	24. Tubinares.	25. Impennes.	26. Gaviæ.	27. Limicolæ.	28. Grallæ.	29. Fulicariæ.	30. Pygopodes.	31. Gallinæ.	32. Crypturi.	33. Apteryges.	34. Rheæ.	35. Casuarii.	36. Struthiones.
A	*	*	*	*	*	*	*	*	*	...	...		*	*	*	*	...	*	...	...	*	*	*	...	...	*	*	*	*	*	*	*	*	*	*	*
B	...	...	...	...	*	...	...	*	...	*	*		*	...	...	*	...	*	*	*	*	*	*	*	*	*	*	...	...	*	*	*	*	*	*	*
C	...	...	...	...	...	...	...	...	...	...	...		...	...	...	*	*	*	*	*	*	*	*	...	*	...	...	...	...	*	...	...	*	*	*	*
D	*	*	*	*	*	...	...	*	*	...	*		*	*	...	*	...	*	*	*	*	...	...	...	*	*	...	...	...	*	...	...	...	...	...	...
E	...	...	...	*	...	*	...	...	...	...	...		...	*	*	*	*	...	...	...	...	...	...	*	...	...	...	...	...	...	...	...	...	*	*	*

A still shorter diagnosis may be made if the muscles of the thigh be allowed to figure as characters. The Owls, the Pelicans, the Humming-Birds, and the Swifts are the only birds which only possess the femoro-caudal muscle of the five included in Garrod's formula. As none of them except the Owls possess basipterygoid processes, the diagnosis may be completed with only two characters.

The Owls are further remarkable for having the basal phalanx of the third digit shortened almost to a cube, a condition probably only found elsewhere in the Cypselidæ, in which amily the basipterygoid processes are absent.

Order **PSITTACI.**

The order Psittaci contains one suborder only.

Suborder 13. PSITTACI.

The Parrots may be diagnosed as follows :—

A. Feet zygodactyle. **B**. Dorsal vertebræ opisthocœlous.

The following combination is also diagnostic :—

C. Young born helpless and nearly naked. **D**. Spinal feather-tract well defined on neck, and forked on upper back. **E**. Deep plantar tendons galline. **F**. Oil-gland tufted or absent.

	1. Passeres.	2. Eurylæmi.	3. Trochili.	4. Scansores.	5. Upupæ.	6. Trogones.	7. Columbæ.	8. Musophagi.	9. Cuculi.	10. Serpentarii.	11. Accipitres.	12. Striges.	13. Psittaci.	14. Halcyones.	15. Coraciæ.	16. Bucerotes.	17. Cathartes.	18. Steganopodes.
A	*	*	*	...	*	*	*	...	...	*	...	...		...	†	*	*	*
B	*	*	*	*	*	*	*	*	*	*	*	*		*	†	*	*	...
C	...	...	...	...	...	...	...	...	...	*	*	*		...	...	...	*	...
D	*	*	*	*	...	*	...	*	...	...	...	...		*	...	*	*	*
E	*	...	...	*	*	*	...	...	...	...	...	...		*	*	*	*	...
F	*	*	*	...	...	*	*	...	*	...	...	*		...	*	...	*	...

	19. Herodiones.	20. Plataleæ.	21. Phœnicopteri.	22. Anseres.	23. Palamedeæ.	24. Tubinares.	25. Impennes.	26. Gaviæ.	27. Limicolæ.	28. Grallæ.	29. Fulicariæ.	30. Pygopodes.	31. Gallinæ.	32. Crypturi.	33. Apteryges.	34. Rheæ.	35. Casuarii.	36. Struthiones.
A	*	*	*	*	*	*	*	*	*	*	*	*	*	*	*	*	*	*
B	*	*	*	*	*	*	...	...	...	*	*	*	*	*	*	*	*	*
C	...	...	*	*	*	*	*	*	*	*	*	*	*	*	*	*	*	*
D	*	*	*	*	*	...	*	...	...	...	...	*	...	*	*	*	*	*
E	...	...	...	...	...	...	...	...	...	...	...	...	...	...	...	...	...	...
F	...	...	...	...	...	...	...	...	...	...	...	...	...	...	...	...	...	...

The subdivision of the Psittaci is a question upon which much difference of opinion exists. Various attempts have been made by Finsch, Reichenow, and others to classify the Parrots on external characters, and at least one ornithologist has tried to make use of internal characters for this purpose (Garrod, Proc. Zool. Soc. 1874, pp. 589–598). The characters are based upon the arrangement of the carotid arteries ; the presence or absence of the ambiens muscle ; the presence or absence of the furcula ; the presence or absence of the oil-gland ; and the completeness or otherwise of the infraorbital septum.

† Character **A** excludes suborder 15, except the Leptosomidæ. Character **B** excludes suborder 15, except the Steatornithidæ.

Subclass CORACIIFORMES.

The Coraciiformes may be diagnosed by a single character. So far as is known, they differ from every other bird in the arrangement of their deep plantar tendons. In all other birds the hallux (if it be present and important enough to have any connection with the deep plantar tendons) is connected with the *flexor longus hallucis*, and not with the *flexor perforans digitorum*. All the Coraciiformes have a hallux, and in all of them it is connected with the *flexor perforans digitorum*, and not with the *flexor longus hallucis*. The two plantars are always coalesced (as they are in the Accipitres, Anseres, &c.), but may easily be separated by gently tearing them asunder, but in very many cases the tendon to the hallux branches off from the *flexor perforans digitorum* before the two plantars coalesce.

It is scarcely possible that such a remarkable condition of the deep plantar tendons could have been independently acquired by two groups of birds.

Order **PICARIÆ.**

The Picariæ or Syndactyli may be diagnosed by their combination of two characters :—

A. Flexor perforans digitorum leading to hallux. **B.** Ambiens muscle absent.

The first character excludes them from every bird except Cathartes, which is excluded by the second character.

Other characters may be added :—

The young are born nearly naked.

Suborder 14. HALCYONES.

The Kingfishers and their allies may be diagnosed in various ways:—

A. Palate desmognathous. **B.** Hallux always present, and connected with the *flexor perforans digitorum.* **C.** Spinal feather-tract well defined on neck, and not forked on the back. **D.** Basipterygoid processes absent.

	1. Passeres.	2. Eurylæmi.	3. Trochili.	4. Scansores.	5. Upupæ.	6. Trogones.	7. Columbæ.	8. Musophagi.	9. Cuculi.	10. Serpentarii.	11. Accipitres.	12. Striges.	13. Psittaci.	14. Halcyones.	15. Coraciæ.	16. Bucerotes.	17. Cathartes.	18. Steganopodes.
A	*	*	*	...	...	*	*	...	...	...	...	...	...		...	...	...	...
B	*	*	*	*	*	*	*	*	*	*	*	*	*		...	...	...	*
C	...	...	...	...	*	...	*	...	*	*	*	*	*		*	*	*	*
D	...	...	...	...	...	*	...	...	...	*	...	*	...		...	...	*	...

	19. Herodiones.	20. Plataleæ.	21. Phœnicopteri.	22. Anseres.	23. Palamedeæ.	24. Tubinares.	25. Impennes.	26. Gaviæ.	27. Limicolæ.	28. Grallæ.	29. Fulicariæ.	30. Pygopodes.	31. Gallinæ.	32. Crypturi.	33. Apteryges.	34. Rheæ.	35. Casuarii.	36. Struthiones.
A	...	...	...	...	...	*	*	*	*	*	*	*	*	*	*	*	*	*
B	*	*	*	*	*	*	*	*	*	*	*	*	*	*	*	*	*	*
C	*	*	*	*	*	*	*	*	*	*	*	*	...	...	*	*	*	*
D	...	...	...	*	*	...	...	...	*	...	...	...	*	*	*	*	*	*

The Halcyones consist of four families:—

Coliidæ.—The Colies are the most aberrant group, and differ from the other three in two important points. The feet are not syndactylous, and there are no bare tracts on the breast (Garrod, Proc. Zool. Soc. 1876, p. 419).

Alcedinidæ.—The Kingfishers agree with the Colies in having a tufted oil-gland and no cæca; but they differ from them, and agree with the Todies and Motmots, in having syndactyle feet, and in having the ventral feather-tract not only split in the centre, but also on each side of the breast by bare tracts. The Kingfishers differ from the other three families in having no accessory semitendinosus.

Momotidæ.—The Motmots agree with the Kingfishers and the Colies in having no cæca, but the tuft of the oil-gland is either absent or very small. The Motmots further differ from the other three families in having the two notches on each side of the posterior margin of the sternum closed into foramina instead of being open.

Todidæ.—The Todies differ in both these particulars from the Motmots. They have large cæca, and well-developed tufts to the oil-gland (Forbes, Proc. Zool. Soc. 1882, p. 448).

Suborder 15. CORACIÆ.

The Goatsuckers, Rollers, and their allies may be diagnosed as follows :—

A. Hallux present, and connected with the *flexor perforans digitorum*. **B.** Spinal feather-tract well defined on the neck, but forked on the upper back.

To this diagnosis may be added other characters :—

C. Young born helpless. **D.** Oil-gland not tufted.

	1. Passeres.	2. Eurylæmi.	3. Trochili.	4. Scansores.	5. Upupæ.	6. Trogones.	7. Columbæ.	8. Musophagi.	9. Cuculi.	10. Serpentarii.	11. Accipitres.	12. Striges.	13. Psittaci.	14. Halcyones.	15. Coraciæ.	16. Bucerotes.	17. Cathartes.	18. Steganopodes.
A	*	*	*	*	*	*	*	*	*	*	*	*	*	...		...	...	*
B	*	*	*	*	...	*	...	*	...	...	...	...	...	*		*	*	*
C	...	...	...	...	...	...	...	...	...	...	...	...	...	...		...	...	...
D	...	...	...	...	*	...	...	*	...	*	*	...	*	...		*	*	*

	19. Herodiones.	20. Plataleæ.	21. Phœnicopteri.	22. Anseres.	23. Palamedeæ.	24. Tubinares.	25. Impennes.	26. Gaviæ.	27. Limicolæ.	28. Grallæ.	29. Fulicariæ.	30. Pygopodes.	31. Gallinæ.	32. Crypturi.	33. Apteryges.	34. Rheæ.	35. Casuarii.	36. Struthiones.
A	*	*	*	*	*	*	*	*	*	*	*	*	*	*	*	*	*	*
B	*	*	*	*	*	...	*	...	...	...	...	*	...	*	*	*	*	*
C	...	...	*	*	*	...	...	*	*	*	*	*	*	*	*	*	*	*
D	*	*	*	*	*	*	*	*	*	...	...	*	...	*	...	...	...	...

The Coraciæ consist of seven families, which are very easily distinguished from each other :—

Cypselidæ.—The Swifts are the only family in the suborder which have no cæca, or which have an ægithognathous palate, or which are without the semitendinosus muscle.

Caprimulgidæ.—The Goatsuckers are the only birds in the suborder which are schizognathous, and they and the Oil-Bird are the only birds in the suborder which have basipterygoid processes.

Steatornithidæ.—The Oil-Bird is the only bird in the suborder which has opisthocœlous dorsal vertebræ, and it and the Goatsuckers are the only birds in the suborder which have basipterygoid processes (Parker, Proc. Zool. Soc. 1889, p. 161).

Podargidæ.—The Frog-mouths are the only family in the suborder which are without an oil-gland, and they and the Madagascar Roller are the only families in the suborder which have a large powder-down patch on each side of the rump.

Leptosomidæ.—The Madagascar Roller and the Frog-mouths differ from the other five families in having a large powder-down patch on each side of the rump. The Madagascar Roller differs from all the other families in being more or less zygodactyle, and in having a horny plate (instead of a membrane) across the nasal opening (Sclater, Proc. Zool. Soc. 1865, p. 682).

Coraciidæ.—The Rollers are the least specialized family in the suborder, and possess none of the characters herein enumerated as peculiar to each of the other six families.

Meropidæ.—The Bee-eaters differ from the other six families in having the episternal process forked (as in most of the Passeres), and perforated to receive the feet of the coracoids (as in the Upupidæ and Bucerotidæ).

Suborder 16. BUCEROTES.

The Hornbills may be diagnosed as follows :—

A. Hallux present, and connected with *flexor perforans digitorum*. **B**. Basipterygoid processes absent. **C**. Spinal feather-tract not defined on the neck. **D**. Episternal process perforated to receive the feet of the coracoids.

Of the above characters the combinations **A B C** and **C D** are each diagnostic.

	1. Passeres.	2. Eurylæmi.	3. Trochili.	4. Scansores.	5. Upupæ.	6. Trogones.	7. Columbæ.	8. Musophagi.	9. Cuculi.	10. Serpentarii.	11. Accipitres.	12. Striges.	13. Psittaci.	14. Halcyones.	15. Coraciæ.	16. Bucerotes.	17. Cathartes.	18. Steganopodes.
A	*	*	*	*	*	*	*	*	*	*	*	*	*	...	...		...	*
B	...	...	...	...	...	*	*	...	...	*	...	*	...	...	...		*	...
C	*	*	*	*	*	*	*	*	*	*	*	*	*	*	*		...	...
D	*	*	*	*	...	*	*	*	*	*	*	*	*	*	...		*	*

	19. Herodiones.	20. Platalеæ.	21. Phœnicopteri.	22. Anseres.	23. Palamedeæ.	24. Tubinares.	25. Impennes.	26. Gaviæ.	27. Limicolæ.	28. Grallæ.	29. Fulicariæ.	30. Pygopodes.	31. Gallinæ.	32. Crypturi.	33. Apteryges.	34. Rheæ.	35. Casuarii.	36. Struthiones.
A	*	*	*	*	*	*	*	*	*	*	*	*	*	*	*	*	*	*
B	...	...	...	*	*	...	...	...	*	...	...	...	*	*	*	*	*	*
C	...	...	...	...	...	*	...	*	...	...	...	...	*	*	...	...	...	...
D	*	*	*	*	*	*	*	*	*	*	*	*	...	*	*	*	*	*

The absence of lateral bare tracts on the neck, which occurs for the first time in the Bucerotes, is a Struthionine character.

The perforation of the episternal process to receive the feet of the coracoids is a character found in the Gallinæ, the Bucerotes, and the Upupæ. It also occurs amongst the Coraciæ in the family Meropidæ.

Order **CATHARTES.**

The order Cathartes contains one suborder only.

Suborder 17. CATHARTES.

The Pseudo-Vultures may be diagnosed in various ways :—

A. Hallux present and connected with the *flexor perforans digitorum.* **B.** Basipterygoid processes present. **C.** Spinal feather-tract not defined on neck. **D.** Young born helpless, but covered with down.

Of the above characters the combinations **A B C** and **B C D** are equally diagnostic.

	1. Passeres.	2. Eurylæmi.	3. Trochili.	4. Scansores.	5. Upupæ.	6. Trogones.	7. Columbæ.	8. Musophagi.	9. Cuculi.	10. Serpentarii.	11. Accipitres.	12. Striges.	13. Psittaci.	14. Halcyones.	15. Coraciæ.	16. Bucerotes.	17. Cathartes.	18. Steganopodes.
A	*	*	*	*	*	*	*	*	*	*	*	*	*	...	...	...		*
B	*	*	*	*	*	...	...	*	*	...	*	...	*	*	...	*		*
C	*	*	*	*	*	*	*	*	*	*	*	*	*	*	*	...		...
D	*	*	*	*	*	*	*	*	*	...	...	...	*	*	...	*		*

	19. Herodiones.	20. Plataleæ.	21. Phœnicopteri.	22. Anseres.	23. Palamedeæ.	24. Tubinares.	25. Impennes.	26. Gaviæ.	27. Limicolæ.	28. Grallæ.	29. Fulicariæ.	30. Pygopodes.	31. Gallinæ.	32. Crypturi.	33. Apteryges.	34. Rheæ.	35. Casuarii.	36. Struthiones.
A	*	*	*	*	*	*	*	*	*	*	*	*	*	*	*	*	*	*
B	*	*	*	...	...	...	*	*	...	...	...	*	...	...	...	...	...	...
C	...	...	...	...	...	*	...	*	...	...	...	...	*	*	...	...	...	...
D	*	...	*	*	*	...	...	*	*	*	*	*	*	*	*	*	*	*

The Cathartes differ from the Vultures of the Old World in many other points besides those specified. The hallux is small and elevated above the level of the other toes; the semitendinosus and accessory semitendinosus muscles are present; the femoro-caudal muscle is sometimes absent; the oil-gland is nude; the maxillo-palatines are much wider apart, and though the distance is bridged over by an ossified nasal septum the ossification does not prevent the nostrils being pierced; they have no cæca &c. &c.

Garrod placed the Cathartes between the Herons and the Storks.

Subclass ANSERIFORMES.

The Anseriformes may be diagnosed by three characters, to which others may be added :—

A. Palate desmognathous. **B.** Spinal feather-tract not defined on the neck (either coalesced with the ventral feather-tracts, or its place occupied by a spinal bare tract). **C.** *Flexor perforans digitorum* never leading to hallux. **D.** Young always passing through a downy stage.

	1. Passeres.	2. Eurylæmi.	3. Trochili.	4. Scansores.	5. Upupæ.	6. Trogones.	7. Columbæ.	8. Musophagi.	9. Cuculi.	10. Serpentarii.	11. Accipitres.	12. Striges.	13. Psittaci.	14. Halcyones.	15. Coraciæ.	16. Bucerotes.	17. Cathartes.	18. Steganopodes.
A	*	*	*	...	...	*	*	...	...	...	...	...	...	...	...	...	...	
B	*	*	*	*	*	*	*	*	*	*	*	*	*	*	*	...	...	
C	...	...	...	...	...	...	...	...	...	...	...	...	...	*	*	*	*	
D	*	*	*	*	*	*	*	*	*	...	...	...	...	*	...	*	...	

	19. Herodiones.	20. Plataleæ.	21. Phœnicopteri.	22. Anseres.	23. Palamedeæ.	24. Tubinares.	25. Impennes.	26. Gaviæ.	27. Limicolæ.	28. Grallæ.	29. Fulicariæ.	30. Pygopodes.	31. Gallinæ.	32. Crypturi.	33. Apteryges.	34. Rheæ.	35. Casuarii.	36. Struthiones.
A						*	*	*	*	*	*	*	*	*	*	*	*	*
B						*	...	*	...	...	...	...	*	*	...	...	...	...
C						...	...	...	...	...	...	...	...	...	...	...	...	...
D						...	...	...	...	...	...	...	...	...	...	...	...	..

The presence of well-marked supraorbital fossæ for the nasal glands is a very important character in determining the position of some birds. These fossæ are specially characteristic of swimming and diving birds, and reach their greatest development in the Impennes, the Tubinares, the Alcidæ, the Pygopodes, &c. The Œdicnemidæ are holorhinal like the Fulicariæ, but have opisthocœlous dorsal vertebræ like the Gaviæ. Into which group must they be placed? Ought they to be holorhinal birds side by side with the schizorhinal Cursoriidæ, or are they better placed as opisthocœlous birds next to the heterocœlous Otididæ? The fact that the Cursoriidæ (like the rest of the Gaviæ) have well-marked supraorbital fossæ, and that the Otididæ (like the rest of the Fulicariæ) are without this character, is a strong argument for placing the Œdicnemidæ in the Gaviæ, especially since they have long ceased to be swimmers or divers.

Order PELECANO-HERODIONES.

The Pelecano-Herodiones may be diagnosed by the following four characters :—

A. Spinal feather-tract not defined on the neck. **B.** *Flexor perforans digitorum* not leading to the hallux. **C.** Young born helpless, but passing through a downy stage. **D.** Palate desmognathous.

	1. Passeres.	2. Eurylæmi.	3. Trochili.	4. Scansores.	5. Upupæ.	6. Trogones.	7. Columbæ.	8. Musophagi.	9. Cuculi.	10. Serpentarii.	11. Accipitres.	12. Striges.	13. Psittaci.	14. Halcyones.	15. Coraciæ.	16. Buceroteѕ.	17. Cathartes.	18. Steganopodes.
A	*	*	*	*	*	*	*	*	*	*	*	*	*	*	*	...	...	
B	...	...	...	...	...	...	...	...	...	...	...	...	...	*	*	*	*	
C	*	*	*	*	*	*	*	*	...	...	...	...	...	*	...	...	*	
D	*	*	*	...	...	*	*	...	...	...	...	...	...	...	...	...	...	

	19. Herodiones.	20. Plataleæ.	21. Phœnicopteri.	22. Anseres.	23. Palamedeæ.	24. Tubinares.	25. Impennes.	26. Gaviæ.	27. Limicolæ.	28. Grallæ.	29. Fulicariæ.	30. Pygopodes.	31. Gallinæ.	32. Crypturi.	33. Apteryges.	34. Rheæ.	35. Casuarii.	36. Struthiones.
A			...	...	...	*	...	*	...	...	...	...	*	*	...	...	...	...
B			...	...	...	...	...	...	...	...	...	...	...	...	...	...	...	...
C			*	*	*	...	...	*	*	*	*	*	*	*	*	*	*	*
D			...	...	...	*	*	*	*	*	*	*	*	*	*	*	*	*

The Pelecano-Herodiones differ from each other very much in their myological formula, which is varied as follows :—A+, A−, AX+, XY+, AXY+, AXY−, and ABXY+ ; nevertheless they were all included by Garrod in his Ciconiiformes.

The femoro-caudal (A) is absent in some species and present in others in the following suborders :—Coraciæ, Cathartes, Herodiones, Gaviæ, Grallæ, Fulicariæ, Pygopodes, Gallinæ, Casuarii, and Struthiones.

The semitendinosus (X) is always present except in the Accipitres, the Striges, the Trochili, in some of the Steganopodes, and some of the Coraciæ.

Whenever the semitendinosus is absent the accessory semitendinosus (Y) is also absent. To the five exceptions above named must be added the Trogones, the Anseres, the Impennes, and some of the Passeres, Scansores, Halcyones, Tubinares, Gaviæ, and Pygopodes.

Suborder 18. STEGANOPODES.

The Pelicans, Cormorants, and their allies are easily diagnosed:—

A. Palate desmognathous. **B.** Sternum not perforated to receive the feet of the coracoids. **C.** Mandible not produced and recurved behind its articulation with the quadrate. **D.** Plumage of neck continuous.

Other characters may be added:—

E. Basipterygoid processes absent.

	1. Passeres.	2. Eurylæmi.	3. Trochili.	4. Scansores.	5. Upupæ.	6. Trogones.	7. Columbæ.	8. Musophagi.	9. Cuculi.	10. Serpentarii.	11. Accipitres.	12. Striges.	13. Psittaci.	14. Halcyones.	15. Coraciæ.	16. Bucerotes.	17. Cathartes.	18. Steganopodes.	19. Herodiones.	20. Plataleæ.	21. Phœnicopteri.	22. Anseres.	23. Palamedeæ.	24. Tubinares.	25. Impennes.	26. Gaviæ.	27. Limicolæ.	28. Grallæ.	29. Fulicariæ.	30. Pygopodes.	31. Gallinæ.	32. Crypturi.	33. Apteryges.	34. Rheæ.	35. Casuarii.	36. Struthiones.
A	*	*	*	...	...	*	*	...	...	...	...	...	...	...	...	...	...		...	...	...	...	...	*	*	*	*	*	*	*	*	*	*	*	*	*
B	...	...	...	...	*	...	...	...	...	...	...	...	...	...	...	*	...		...	...	...	...	...	...	...	...	...	...	...	...	*	...	...	...	...	...
C	...	...	...	...	...	...	...	...	...	...	...	...	...	...	...	...	...		...	*	*	*	*	...	...	...	...	...	...	...	*	...	...	...	...	...
D	*	*	*	*	*	*	*	*	*	*	*	*	*	*	*	...	...		*	...	...	...	...	*	...	*	...	...	...	...	*	*	...	...	...	...
E	...	...	...	...	...	*	*	...	...	*	...	*	...	...	...	...	*		...	...	...	*	*	...	...	...	*	...	...	...	*	*	*	*	*	*

The Steganopodes include the following families:—Sulidæ, Phalacrocoracidæ, Pelecanidæ, Phaetontidæ, and Fregatidæ, which are very easily diagnosed by excellent osteological characters (Seebohm, 'Ibis,' 1889, p. 102).

Fregatidæ.—Clavicle ankylosed to sternum; no complete carotid arches on cervical vertebræ.

Phaetontidæ.—Nasal apertures large; palatines not coalesced.

Pelecanidæ.—Clavicle ankylosed to sternum; dorsal vertebræ heterocœlous.

Sulidæ.—Clavicle not ankylosed to sternum; dorsal vertebræ without ventral processes.

Phalacrocoracidæ.—Dorsal vertebræ with ventral processes, and opisthocœlous.

Suborder 19. HERODIONES.

The Storks and the Herons may be diagnosed as follows :—

A. Palate desmognathous. **B.** Spinal bare tract reaching far up the neck. **C.** Mandible not produced and recurved behind its articulation with the quadrate.

This diagnosis may be strengthened by many other characters :—

D. Basipterygoid processes absent. **E.** Young born helpless.

	1. Passeres.	2. Eurylæmi.	3. Trochili.	4. Scansores.	5. Upupæ.	6. Trogones.	7. Columbæ.	8. Musophagi.	9. Cuculi.	10. Serpentarii.	11. Accipitres.	12. Striges.	13. Psittaci.	14. Halcyones.	15. Coraciæ.	16. Bucerotes.	17. Cathartes.	18. Steganopodes.	19. Herodiones.	20. Plataleæ.	21. Phœnicopteri.	22. Anseres.	23. Palamedeæ.	24. Tubinares.	25. Impennes.	26. Gaviæ.	27. Limicolæ.	28. Grallæ.	29. Fulicariæ.	30. Pygopodes.	31. Gallinæ.	32. Crypturi.	33. Apteryges.	34. Rheæ.	35. Casuarii.	36. Struthiones.
A	*	*	*	...	...	*	*	...	...	...	...	...	...	...	...	...	...	...		...	...	...	...	*	*	*	*	*	*	*	*	*	*	*	*	*
B	*	*	*	*	*	*	*	*	*	*	*	*	*	*	*	*	*	*		...	...	...	*	*	*	*	*	...	*	*	*	*	*	*	*	*
C	...	...	...	...	...	...	...	...	...	...	...	...	...	...	...	...	...	...		*	*	*	*	...	...	...	...	...	...	...	*	...	...	...	...	...
D	...	...	...	...	...	*	*	...	...	*	...	*	...	...	...	...	*	...		...	...	*	*	...	...	...	*	...	...	...	*	*	*	*	*	*
E	...	...	...	...	...	...	...	...	...	...	...	...	...	...	...	...	...	...		...	*	*	*	...	...	*	*	*	*	*	*	*	*	*	*	*

The Herodiones consist of three families :—Ardeidæ, Scopidæ, and Ciconiidæ.

Ardeidæ.—Interclavicle projecting conspicuously within the angle of the furculum; a large powder-down patch on each side of the rump; no lateral bare tracts on neck.

Scopidæ.—No interclavicular process within the angle of the furculum; no powder-down patches; well-defined lateral bare tracts on neck.

Ciconiidæ.—No interclavicular process within the angle of the furculum; no powder-down patches; no lateral bare tracts on neck.

Suborder 20. PLATALEÆ.

The Ibises and Spoonbills may be diagnosed as follows :—

A. Palate desmognathous. **B.** Nasals schizorhinal.

This diagnosis may be strengthened by many other characters :—

C. Basipterygoid processes absent. **D.** Spinal feather-tract not defined on the neck.

	1. Passeres.	2. Eurylæmi.	3. Trochili.	4. Scansores.	5. Upupæ.	6. Trogones.	7. Columbæ.	8. Musophagi.	9. Cuculi.	10. Serpentarii.	11. Accipitres.	12. Striges.	13. Psittaci.	14. Halcyones.	15. Coraciæ.	16. Bucerotes.	17. Cathartes.	18. Steganopodes.	19. Herodiones.	20. Plataleæ.	21. Phœnicopteri.	22. Anseres.	23. Palamedeæ.	24. Tubinares.	25. Impennes.	26. Gaviæ.	27. Limicolæ.	28. Grallæ.	29. Fulicariæ.	30. Pygopodes.	31. Gallinæ.	32. Crypturi.	33. Apteryges.	34. Rheæ.	35. Casuarii.	36. Struthiones.
A	*	*	*	...	...	*	*	...	...	...	...	...	...	...	...	...	...	...	...		...	...	...	*	*	*	*	*	*	*	*	*	*	*	*	*
B	...	*	*	*	*	*	...	*	*	*	*	*	*	*	*	*	*	*	*		*	*	*	*	*	...	...	...	*	*	*	...	*	*	*	*
C	...	...	...	...	...	*	*	...	...	*	...	*	...	...	...	...	*	...	...		...	*	*	...	...	...	*	...	...	...	*	*	*	*	*	*
D	*	*	*	*	*	*	*	*	*	*	*	*	*	*	*	...	...	...	...		...	...	...	*	...	*	...	...	...	...	*	*	...	...	...	...

It is possible that characters of sufficient importance may be found to separate the Ibididæ from the Plataleidæ.

Garrod and Forbes placed *Platalea* and *Ibis* amongst the Charadriiformes and not amongst the Anseriformes, because they regarded the schizorhinal bifurcation of the nasals as of greater taxonomic value than the desmognathous character of the palate. The discovery of some small genera of schizorhinal birds amongst the otherwise holorhinal Passeres did not alter their opinion. I should much have preferred to degrade the Plataleæ to a family of the Herodiones, but the necessity of diagnosing the groups prevented me from following my wishes. It is very easy to diagnose the Herodiones, and still easier to diagnose the Plataleæ, but all my attempts to diagnose the group which combined the two proved failures.

Order **LAMELLIROSTRES.**

So far as is known, the Lamellirostres are the only birds which combine the following two characters :—

A. Palate desmognathous. **B.** Young born covered with down, and able to run in a few hours.

	1. Passeres.	2. Eurylæmi.	3. Trochili.	4. Scansores.	5. Upupæ.	6. Trogones.	7. Columbæ.	8. Musophagi.	9. Cuculi.	10. Serpentarii.	11. Accipitres.	12. Striges.	13. Psittaci.	14. Halcyones.	15. Coraciæ.	16. Bucerotes.	17. Cathartes.	18. Steganopodes.
A	*	*	*	...	...	*	*	...	...	...	...	...	...	...	...	...	...	...
B	*	*	*	*	*	*	*	*	*	*	*	*	*	*	*	*	*	*

	19. Herodiones.	20. Plataleæ.	21. Phœnicopteri.	22. Anseres.	23. Palamedeæ.	24. Tubinares.	25. Impennes.	26. Gaviæ.	27. Limicolæ.	28. Grallæ.	29. Fulicariæ.	30. Pygopodes.	31. Gallinæ.	32. Crypturi.	33. Apteryges.	34. Rheæ.	35. Casuarii.	36. Struthiones.
A	...	..				*	*	*	*	*	*	*	*	*	*	*	*	*
B	*	*				*	*	...	...	...	...	...	...	...	...	...	...	...

Some doubt still lingers over the history of the Flamingoes. Disputes have arisen as to their affinities, some ornithologists regarding them as allied to the Ducks, whilst others have supposed them to be more nearly related to the Storks. The mystery which so long enveloped their breeding-habits has not yet been entirely dissolved. There is much circumstantial evidence in favour of the theory that the young are born covered with down and able to run about in a few hours, like young Ducks, and that they are neither born naked like young Herons, nor require to be fed in the nest by their parents for many days like young Petrels; but until their breeding-habits have been accurately observed and described it is impossible not to feel some doubt as to their real position.

Suborder 21. PHŒNICOPTERI.

The Flamingoes may be diagnosed as follows :—

A. Palate desmognathous. **B.** Basipterygoid processes absent or very rudimentary. **C.** Nasals holorhinal. **D.** Mandible much produced and recurved behind its articulation with the quadrate.

	1. Passeres.	2. Eurylæmi.	3. Trochili.	4. Scansores.	5. Upupæ.	6. Trogones.	7. Columbæ.	8. Musophagi.	9. Cuculi.	10. Serpentarii.	11. Accipitres.	12. Striges.	13. Psittaci.	14. Halcyones.	15. Coraciæ.	16. Bucerotes.	17. Cathartes.	18. Steganopodes.
A	*	*	*	...	...	*	*	...	...	...	...	...	...	...	...	...	...	...
B	...	...	...	...	...	*	*	...	...	*	...	*	...	...	...	...	*	...
C	...	...	...	...	...	...	*	...	...	...	...	...	...	...	...	...	...	...
D		*	*	*	*	*	*	*	*	*	*	*	*	*	*	*	*	*

	19. Herodiones.	20. Plataleæ.	21. Phœnicopteri.	22. Anseres.	23. Palamedeæ.	24. Tubinares.	25. Impennes.	26. Gaviæ.	27. Limicolæ.	28. Grallæ.	29. Fulicariæ.	30. Pygopodes.	31. Gallinæ.	32. Crypturi.	33. Apteryges.	34. Rheæ.	35. Casuarii.	36. Struthiones.
A	...	...		...	...	*	*	*	*	*	*	*	*	*	*	*	*	*
B	...	...		*	*	...	...	...	*	...	...	...	*	*	*	*	*	*
C	...	*		...	...	...	...	...	*	*	...	...	...	*	...	...	...	...
D	*	...		...	...	*	*	*	...	*	*	*	...	*	*	*	*	*

The presence or absence of the hallux is a character of some importance, though it rarely denotes relationship. It may be assumed that all birds had originally four toes, and that the hallux has been independently lost in many families. In the Tubinares the hallux is either absent or is only represented by one phalanx. The Struthiones, the Casuarii, and the Rheæ have no hallux. The following suborders contain some species which have also lost the hallux :—Scansores, Phœnicopteri, Gaviæ, Limicolæ, Grallæ, Fulicariæ, and Crypturi.

Suborder 22. ANSERES.

The Ducks, the Geese, and the Swans only require two characters for their diagnosis :—

A. Basipterygoid processes articulating with the pterygoids as near the palatines as possible. **B.** Maxillo-palatines completely coalesced across the middle line.

This diagnosis may be strengthened by the addition of many other characters :—

C. Oil-gland tufted. **D.** Young born covered with down and able to run in a few hours.

	1. Passeres.	2. Eurylæmi.	3. Trochili.	4. Scansores.	5. Upupæ.	6. Trogones.	7. Columbæ.	8. Musophagi.	9. Cuculi.	10. Serpentarii.	11. Accipitres.	12. Striges.	13. Psittaci.	14. Halcyones.	15. Coraciæ.	16. Bucerotes.	17. Cathartes.	18. Steganopodes.
A	*	*	*	*	*	*	*	*	*	*	*	*	*	*	*	*	*	*
B	*	*	*	...	...	*	*	...	...	...	...	...	...	...	...	...	...	...
C	*	*	*	...	...	*	*	...	*	...	...	*	...	...	*	...	*	...
D	*	*	*	*	*	*	*	*	*	*	*	*	*	*	*	*	*	*

	19. Herodiones.	20. Plataleæ.	21. Phœnicopteri.	22. Anseres.	23. Palamedeæ.	24. Tubinares.	25. Impennes.	26. Gaviæ.	27. Limicolæ.	28. Grallæ.	29. Fulicariæ.	30. Pygopodes.	31. Gallinæ.	32. Crypturi.	33. Apteryges.	34. Rheæ.	35. Casuarii.	36. Struthiones.
A	*	*	*		*	*	*	*	*	*	*	*	...	*	*	*	*	*
B	...	...	...		...	*	*	*	*	*	*	*	*	*	*	*	*	*
C	...	...	...		...	...	...	...	...	...	...	...	...	...	*	*	*	*
D	*	*	...		...	*	*	...	...	...	...	...	...	...	...	...	...	...

The Swans appear to belong to a different family to the Ducks and Geese. The Cygnidæ have from 23 to 25 cervical (including cervico-dorsal) vertebræ, whilst the Anatidæ have only from 16 to 19.

Suborder 23. PALAMEDEÆ.

The Screamers are supposed to be absolutely unique amongst existing birds in having lost the uncinate processes of their ribs.

They may otherwise be diagnosed in various ways:—

A. Plumage of upper parts with no spinal bare tract. **B**. Palate desmognathous. **C**. Young born covered with down and able to run in a few hours.

To this diagnosis many other characters may be added:—

D. Number of cervical vertebræ more than 18.

	1. Passeres.	2. Eurylæmi.	3. Trochili.	4. Scansores.	5. Upupæ.	6. Trogones.	7. Columbæ.	8. Musophagi.	9. Cuculi.	10. Serpentarii.	11. Accipitres.	12. Striges.	13. Psittaci.	14. Halcyones.	15. Coraciæ.	16. Bucerotes.	17. Cathartes.	18. Steganopodes.	19. Herodiones.	20. Plataleæ.	21. Phœnicopteri.	22. Anseres.	23. Palamedeæ.	24. Tubinares.	25. Impennes.	26. Gaviæ.	27. Limicolæ.	28. Grallæ.	29. Fulicariæ.	30. Pygopodes.	31. Gallinæ.	32. Crypturi.	33. Apteryges.	34. Rheæ.	35. Casuarii.	36. Struthiones.
A	...	*	*	*	*	...	*	...	*	*	*	*	*	...	*	...	*	...	*	...	*	*		*	*	*	*	*	*	...	...	...	...	...	...	...
B	*	*	*	...	...	*	*	...	...	...	...	...	...	...	...	...	...	...	...	...	...	...		*	*	*	*	*	*	*	*	*	*	*	*	*
C	*	*	*	*	*	*	*	*	*	*	*	*	*	*	*	*	*	*	*	*	...	...		*	*	...	...	...	...	...	...	...	...	...	...	...
D	*	*	*	*	*	*	*	*	*	*	*	*	*	*	*	*	*	...	...	*	...	...		*	*	*	*	...	...	...	*	*	*	*	...	...

In having no uncinate processes to the ribs the Palamedeæ are in advance of all other birds. The toothed birds of North America (*Hesperornis* and *Ichthyornis*) had uncinate processes to their ribs, and it is not unreasonable to assume that all birds once possessed them, and consequently that the Palamedeæ have lost them. Uncinate processes are characteristic of modern Crocodilia, and there can be little doubt that Crocodiles and Birds inherited them from common ancestors. In their pterylosis, which is characterized by the almost entire absence of bare tracts, the Palamedeæ are as archaic as the Ratitæ.

Subclass GALLIFORMES.

The Galliformes are believed to be the only birds which combine the following four characters:—

A. Young born covered with down or feathers. **B**. Maxillo-palatines not united across the middle line. **C**. Keel of sternum well developed. **D**. If a deep plantar tendon reach the hallux it proceeds from the *flexor longus hallucis*, and not from the *flexor perforans digitorum.*

	1. Passeres.	2. Eurylæmi.	3. Trochili.	4. Scansores.	5. Upupæ.	6. Trogones.	7. Columbæ.	8. Musophagi.	9. Cuculi.	10. Serpentarii.	11. Accipitres.	12. Striges.	13. Psittaci.	14. Halcyones.	15. Coraciæ.	16. Bucerotes.	17. Cathartes.	18. Steganopodes.
A	*	*	*	*	*	*	*	*	*	...	...	...	*	*	...	*	...	*
B	...	...	...	...	*	...	...	*	*	*	*	*	*	*	...	*	*	*
C	...	...	...	...	...	...	...	...	...	...	...	...	...	...	...	...	...	...
D	...	...	...	...	...	...	...	...	...	...	...	...	...	*	*	*	*	...

	19. Herodiones.	20. Plataleæ.	21. Phœnicopteri.	22. Anseres.	23. Palamedeæ.	24. Tubinares.	25. Impennes.	26. Gaviæ.	27. Limicolæ.	28. Grallæ.	29. Fulicariæ.	30. Pygopodes.	31. Gallinæ.	32. Crypturi.	33. Apteryges.	34. Rheæ.	35. Casuarii.	36. Struthiones.
A	...	...	...	...	...										...	...	...	...
B	*	*	*	*	*										...	...	...	...
C	...	...	...	...	...										*	*	*	*
D	...	...	...	...	...										...	...	...	..

Some ornithologists (Fürbringer, Morph. und Syst. Vögel, p. 1565) place the Pygopodes between the Anseres and the Herodiones in an Order Pelargornithes, and regard the Tubinares and Impennes as intermediate suborders which connect the Pelargornithes with the Charadriiornithes, but which have an equal claim to be included in either, and must consequently be rejected from both. The Pygopodes appear to me to be highly specialized Fulicariæ, and I have been obliged to make use of the absence of some of their special characters to exclude them from that suborder. The Tubinares may be highly specialized and advanced Gaviæ, and the Impennes are possibly highly specialized and retrograde Gaviæ, but the evidence in favour of these hypotheses is very weak.

Order **TUBINARES.**

Suborder 24. TUBINARES.

The Tubinares are the only birds which possess the following character :—

A. External nostrils produced into tubes.

They are also the only birds which combine the following four characters :—

B. Nasals holorhinal. **C.** Dorsal vertebræ heterocœlous. **D.** Hallux absent or reduced to one phalanx, the other toes directed forwards. **E.** Young fed by the parents for some time in the nest.

To these characters may be added the following :—

F. Spinal feather-tract well defined on the neck. **G.** Oil-gland tufted.

	1. Passeres.	2. Eurylæmi.	3. Trochili.	4. Scansores.	5. Upupæ.	6. Trogones.	7. Columbæ.	8. Musophagi.	9. Cuculi.	10. Serpentarii.	11. Accipitres.	12. Striges.	13. Psittaci.	14. Halcyones.	15. Coraciæ.	16. Bucerotes.	17. Cathartes.	18. Steganopodes.	19. Herodiones.	20. Plataleæ.	21. Phœnicopteri.	22. Anseres.	23. Palamedeæ.	24. Tubinares.	25. Impennes.	26. Gaviæ.	27. Limicolæ.	28. Grallæ.	29. Fulicariæ.	30. Pygopodes.	31. Gallinæ.	32. Crypturi.	33. Apteryges.	34. Rheæ.	35. Casuarii.	36. Struthiones.
A	*	*	*	*	*	*	*	*	*	*	*	*	*	*	*	*	*	*	*	*	*	*	*		*	*	*	*	*	*	*	*	*	*	*	*
B	...	...	...	...	...	...	*	...	...	...	...	...	...	...	...	...	...	...	...	*	...	...	...		...	...	*	*	...	...	...	*	...	...	...	...
C	...	...	...	...	...	...	...	...	...	...	...	...	*	...	...	...	...	...	...	...	...	...	...		*	*	*	...	...	...	...	...	...	...	...	...
D	*	*	*	*	*	*	*	*	*	*	*	*	*	*	*	*	*	*	*	*	...	*	*		*	..	...	...	...	*	*	...	*	...	...	...
E	...	...	...	...	...	...	...	...	...	...	...	...	...	...	...	...	...	...	...	...	*	*	*		...	*	*	*	*	*	*	*	*	*	*	*
F	...	...	...	...	...	...	...	...	...	...	...	...	...	...	...	*	*	*	*	*	*	*	*		*	...	...	...	...	*	..	...	*	*	*	*
G	*	*	*	...	...	*	*	...	*	...	...	*	...	...	*	...	*	...	...	...	...	...	...		...	...	...	...	...	...	...	...	*	*	*	*

The Tubinares include three families :—Diomedeidæ, Puffinidæ, and Procellariidæ :—

Diomedeidæ.—Nasal tubes separated from each other ; no basipterygoid processes.

Puffinidæ.—Nasal tubes side by side ; basipterygoid processes present.

Procellariidæ.—Nasal tubes side by side ; no basipterygoid processes.

Order IMPENNES.

Suborder 25. IMPENNES.

The Penguins possess at least five characters, each of which is diagnostic.

The first digit of the manus is fused with the second in the adult.

The three metatarsal bones of the tarso-metatarsus are very short, and are separated from each other throughout their whole length by deep grooves.

The bones of the forearm are all flattened.

The scapula is very broad, not differing very much in size from the keel of the sternum.

None of the feathers of the wing are differentiated into quills.

The Penguins may also be diagnosed by their combination of the following three characters:—

A. The young are born helpless, but covered with down. **B.** Spinal feather-tract not defined on the neck. **C.** Palate schizognathous.

	1. Passeres.	2. Eurylæmi.	3. Trochili.	4. Scansores.	5. Upupæ.	6. Trogones.	7. Columbæ.	8. Musophagi.	9. Cuculi.	10. Serpentarii.	11. Accipitres.	12. Striges.	13. Psittaci.	14. Halcyones.	15. Coraciæ.	16. Bucerotes.	17. Cathartes.	18. Steganopodes.
A	*	*	*	*	*	*	*	*	*	...	...	...	*	*	*	*	...	*
B	*	*	*	*	*	*	*	*	*	*	*	*	*	*	*	...	...	...
C	*	*	...	...	*	...	...	*	*	*	*	*	*	*	...	*	*	*

	19. Herodiones.	20. Plataleæ.	21. Phœnicopteri.	22. Anseres.	23. Palamedeæ.	24. Tubinares.	25. Impennes.	26. Gaviæ.	27. Limicolæ.	28. Grallæ.	29. Fulicariæ.	30. Pygopodes.	31. Gallinæ.	32. Crypturi.	33. Apteryges.	34. Rheæ.	35. Casuarii.	36. Struthiones.
A	...	...	*	*	*	...		*	*	*	*	*	*	*	*	*	*	*
B	...	...	...	...	...	*		*	...	...	...	...	*	*	...	...	...	...
C	*	*	*	*	*	...		...	...	...	...	...	...	*	*	*	*	*

The Penguins appear to represent, in the Antarctic Ocean, the Puffins and the Auks which abound in the Arctic Ocean. It is difficult to believe that they are not distantly related. The bones of the palate are so very similar in the Impennes, the Pygopodes, and the Alcidæ, that it is difficult to regard the Penguins as far removed from the Gallo-Grallæ. It is, however, very probable that all birds were once schizognathous, and that the coalescence of the maxillo-palatines is a character which has been independently assumed by several groups. In the articulation of their dorsal vertebræ, in their pterylosis, and in many other characters the Impennes are very archaic, but we may regard their abnormal wings as being highly specialized for the purpose of diving, rather than as resembling those of the Apteryx, which appear to be simply degraded by disuse.

Order GALLO-GRALLÆ.

The birds comprised in the order Gallo-Grallæ possess three characters which are not known to be combined in any bird outside the order :—

A. Palate not desmognathous. **B.** Young born covered with down and able to run in a few hours. **C.** Keel of sternum well developed.

	1. Passeres.	2. Eurylæmi.	3. Trochili.	4. Scansores.	5. Upupæ.	6. Trogones.	7. Columbæ.	8. Musophagi.	9. Cuculi.	10. Serpentarii.	11. Accipitres.	12. Striges.	13. Psittaci.	14. Halcyones.	15. Coraciæ.	16. Bucerotes.	17. Cathartes.	18. Steganopodes.
A	...	...	...	...	*	...	...	*	*	*	*	*	*	*	...	...	*	*
B	*	*	*	*	*	*	*	*	*	*	*	*	*	*	*	*	*	*
C	...	...	...	...	...	...	...	...	...	...	...	...	...	...	...	...	...	...

	19. Herodiones.	20. Plataleæ.	21. Phœnicopteri.	22. Anseres.	23. Palamedeæ.	24. Tubinares.	25. Impennes.	26. Gaviæ.	27. Limicolæ.	28. Grallæ.	29. Fulicariæ.	30. Pygopodes.	31. Gallinæ.	32. Crypturi.	33. Apteryges.	34. Rheæ.	35. Casuarii.	36. Struthiones.
A	*	*	*	*	*	...	...								...	...	...	...
B	*	*	...	...	...	*	*								...	...	...	...
C	...	...	...	...	...	...	...								*	*	*	*

In dividing the Gallo-Grallæ into suborders several difficulties present themselves. Of these perhaps the separation of the Grallæ from the Fulicariæ is the most open to dispute. Combined they include twelve families, of which six are schizorhinal and six are holorhinal. If the twelve families be separated according to their pterylosis, six of them have the lateral bare tracts extending some way up the neck, whilst in the other six they scarcely reach the neck at all. There are three courses open to the morphologist. The Grallæ can be divided into four suborders, or they can be divided into two, either upon their pterylosis, or upon their nasals. I have chosen the last-mentioned course, because three of the twelve families possess the remarkable character of having a powder-down patch on each side of the rump. Either the first or the second course would have placed one of these families in a different suborder from that which contained the other two; whilst the last course leaves these three families in one suborder.

Suborder 26. GAVIÆ.

The Gulls and their allies may be diagnosed as follows :—

A. Palate schizognathous. **B.** Dorsal vertebræ more or less opisthocœlous. **C.** Basi-pterygoid processes absent. **D.** Spinal feather-tract forked on the upper back. **E.** Young born covered with down, and able to run in a few hours.

	1. Passeres.	2. Eurylæmi.	3. Trochili.	4. Scansores.	5. Upupæ.	6. Trogones.	7. Columbæ.	8. Musophagi.	9. Cuculi.	10. Serpentarii.	11. Accipitres.	12. Striges.	13. Psittaci.	14. Halcyones.	15. Coraciæ.	16. Bucerotes.	17. Cathartes.	18. Steganopodes.
A	*	*	...	...	*	...	...	*	*	*	*	*	*	*	†	...	*	*
B	*	*	*	*	*	*	*	*	*	*	*	*	...	*	†	*	*	...
C	...	...	...	...	...	*	*	...	...	*	...	*	...	...	...	...	*	...
D	*	*	*	*	...	*	...	*	...	...	...	...	...	*	...	...	...	...
E	*	*	*	*	*	*	*	*	*	*	*	*	*	*	*	*	*	*

	19. Herodiones.	20. Plataleæ.	21. Phœnicopteri.	22. Anseres.	23. Palamedeæ.	24. Tubinares.	25. Impennes.	26. Gaviæ.	27. Limicolæ.	28. Grallæ.	29. Fulicariæ.	30. Pygopodes.	31. Gallinæ.	32. Crypturi.	33. Apteryges.	34. Rheæ.	35. Casuarii.	36. Struthiones.
A	*	*	*	*	*	...	...		...	...	...	...	...	*	*	*	*	*
B	*	*	*	*	*	*	...		...	*	*	*	*	*	*	*	*	*
C	...	...	...	*	*	...	...		*	...	...	...	*	*	*	*	*	*
D	...	...	...	...	...	...	*		...	...	...	...	...	*	*	*	*	*
E	*	*	...	...	...	*	*		...	...	...	...	...	...	...	...	...	...

The Gaviæ include the Alcidæ, the Laridæ, the Cursoriidæ (*Cursorius*, *Thinocorus*, *Attagis*, *Chionis*, *Dromas*, and *Glareola*), and the Œdicnemidæ (*Œdicnemus* and *Pluvianus*).

The Œdicnemidæ are holorhinal; the other families are schizorhinal. The Alcidæ have lateral occipital fontanelles; the Cursoriidæ, like the Œdicnemidæ, are not web-footed, and doubtless possess other good characters to distinguish them from the Laridæ.

† Character **B** excludes all the families in suborder 15, except the Steatornithidæ, which are excluded by character **A**, so that the diagnosis is complete without character **E**.

Suborder 27. LIMICOLÆ.

The Plovers and Snipes really belong to the same group as the Gulls and their allies, but for convenience they may be allowed to form a separate section characterized by the possession of basipterygoid processes. The Limicolæ may be diagnosed as follows :—

A. Palate schizognathous. **B.** Dorsal vertebræ opisthocœlous. **C.** Basipterygoid processes present.

To this diagnosis many other characters may be added :—

D. Spinal feather-tract forked on the upper back.

	1. Passeres.	2. Eurylæmi.	3. Trochili.	4. Scansores.	5. Upupæ.	6. Trogones.	7. Columbæ.	8. Musophagi.	9. Cuculi.	10. Serpentarii.	11. Accipitres.	12. Striges.	13. Psittaci.	14. Halcyones.	15. Coraciæ.	16. Bucerotes.	17. Cathartes.	18. Steganopodes.	19. Herodiones.	20. Plataleæ.	21. Phœnicopteri.	22. Anseres.	23. Palamedeæ.	24. Tubinares.	25. Impennes.	26. Gaviæ.	27. Limicolæ.	28. Grallæ.	29. Fulicariæ.	30. Pygopodes.	31. Gallinæ.	32. Crypturi.	33. Apteryges.	34. Rheæ.	35. Casuarii.	36. Struthiones.
A	*	*	...	...	*	...	...	*	*	*	*	*	*	*	†	...	*	*	*	*	*	*	*	...	...	...		...	...	...	...	*	*	*	*	*
B	*	*	*	*	*	*	*	*	*	*	*	*	...	*	†	*	*	...	*	*	*	*	*	*	...	...		*	*	*	*	*	*	*	*	*
C	*	*	*	*	*	...	...	*	*	...	*	...	*	*	...	*	...	*	*	*	*	...	...	*	*	*		...	...	*	...	...	...	...	...	...
D	*	*	*	*	...	*	...	*	...	...	...	...	...	*	...	...	...	...	...	...	...	...	...	...	*	...		...	...	...	...	*	*	*	*	*

The Limicolæ include the Charadriidæ and the Parridæ.

The Charadriidæ have lateral occipital fontanelles, which are not found in the Parridæ.

† Character **B** excludes all the families in suborder 15, except the Steatornithidæ, which are excluded by character **A.**

Suborder 28. GRALLÆ.

The Cranes and their allies may be diagnosed as follows :—

A. Maxillo-palatines not coalesced with each other or with the vomer. **B**. Nasals schizorhinal. **C**. Dorsal vertebræ heterocœlous. **D**. Young born covered with down, and able to run in a few hours. **E**. If a deep plantar tendon reach the hallux, it proceeds from the *flexor longus hallucis* and not from the *flexor perforans digitorum*.

	1. Passeres.	2. Eurylæmi.	3. Trochili.	4. Scansores.	5. Upupæ.	6. Trogones.	7. Columbæ.	8. Musophagi.	9. Cuculi.	10. Serpentarii.	11. Accipitres.	12. Striges.	13. Psittaci.	14. Halcyones.	15. Coraciæ.	16. Bucerotes.	17. Cathartes.	18. Steganopodes.
A	...	...	...	...	*	...	...	*	*	*	*	*	*	*	...	*	*	*
B	...	*	*	*	*	*	...	*	*	*	*	*	*	*	*	*	*	*
C	...	...	...	...	...	...	...	...	...	...	...	...	*	...	...	...	...	...
D	*	*	*	*	*	*	*	*	*	*	*	*	*	*	*	*	*	*
E	...	...	...	...	...	...	...	...	...	...	...	...	...	*	*	*	*	...

	19. Herodiones.	20. Plataleæ.	21. Phœnicopteri.	22. Anseres.	23. Palamedeæ.	24. Tubinares.	25. Impennes.	26. Gaviæ.	27. Limicolæ.	28. Grallæ.	29. Fulicariæ.	30. Pygopodes.	31. Gallinæ.	32. Crypturi.	33. Apteryges.	34. Rheæ.	35. Casuarii.	36. Struthiones.
A	*	*	*	*	*	...	...	...	...		...	...	...	*	*	*	*	*
B	*	...	*	*	*	*	*	...	...		*	*	*	...	*	*	*	*
C	...	...	...	...	...	...	*	*	*		...	...	...	...	...	...	...	...
D	*	*	...	...	...	*	*	...	...		...	...	...	...	...	...	...	...
E	...	...	...	...	...	...	...	...	...		...	...	...	...	...	...	...	...

The Grallæ consist of six families, the two smallest containing only a single species each, and the largest not containing a score species :—

Gruidæ (*Grus*, *Aramus*, &c., Garrod, Proc. Zool. Soc. 1876, p. 275).—The Cranes have no notches on the posterior margin of the sternum ; no powder-down patches on each side of the rump ; a tufted oil-gland ; and the four bare tracts on the body extend some way up the neck.

Turnicidæ (Huxley, Proc. Zool. Soc. 1868, p. 303).—The Hemipodes have a deep notch on each side of the posterior margin of the sternum ; no powder-down patches on each side of the rump ; a tufted oil-gland ; but the spinal bare tract does not reach the neck, though the other three do.

Rhinochetidæ (Parker, Trans. Zool. Soc. vi. p. 501).—The Kagu has no notch on the posterior margin of the sternum ; a nude oil-gland ; a powder-down patch on each side of the rump, and the four bare tracts on the body extend some way up the neck.

Mesitidæ (Forbes, Proc. Zool. Soc. 1882, p. 267).—The Madagascar Kagu has the sternum of a Hemipode, combined with the other characters of the Kagu.

Eurypygidæ.—The Sun-Bitterns have a notch on each side of the posterior margin of the sternum ; a nude oil-gland ; and powder-down patches, but none of the bare tracts on the body extend beyond the base of the neck.

Pteroclidæ.—The Sand-Grouse have two notches on each side of the posterior margin of the sternum ; a nude oil-gland ; and no powder-down patches ; nor any lateral bare tracts on the neck.

Suborder 29. FULICARIÆ.

The Rails and their allies really belong to the same group as the Cranes and their allies, but for convenience they may be allowed to form a separate section characterized by their holorhinal nasals. The Holorhinal Grallæ or Fulicariæ may be diagnosed as follows :—

A. Palate schizognathous. **B.** Young born covered with down, and able to run in a few hours. **C.** Nasals holorhinal. **D.** Dorsal vertebræ heterocœlous. **E.** Episternal process not perforated to receive the feet of the coracoids. **F.** Posterior processes of the ilia separated sufficiently to show a broad sacrum.

	1. Passeres.	2. Eurylæmi.	3. Trochili.	4. Scansores.	5. Upupæ.	6. Trogones.	7. Columbæ.	8. Musophagi.	9. Cuculi.	10. Serpentarii.	11. Accipitres.	12. Striges.	13. Psittaci.	14. Halcyones.	15. Coraciæ.	16. Bucerotes.	17. Cathartes.	18. Steganopodes.	19. Herodiones.	20. Plataleæ.	21. Phœnicopteri.	22. Anseres.	23. Palamedeæ.	24. Tubinares.	25. Impennes.	26. Gaviæ.	27. Limicolæ.	28. Grallæ.	29. Fulicariæ.	30. Pygopodes.	31. Gallinæ.	32. Crypturi.	33. Apteryges.	34. Rheæ.	35. Casuarii.	36. Struthiones.
A	*	*	...	...	*	...	...	*	*	*	*	*	*	*	...	...	*	*	*	*	*	*	*	...	...	...	...	...		...	...	*	*	*	*	*
B	*	*	*	*	*	*	*	*	*	*	*	*	*	*	*	*	*	*	*	*	...	...	...	*	*	...	...	...		...	...	...	...	...	...	...
C	...	...	...	...	...	...	*	...	...	...	...	...	...	...	...	...	...	...	...	*	...	...	...	...	...	...	*	*		...	...	*	...	...	...	...
D	...	...	...	...	...	...	...	...	...	...	...	...	*	...	...	...	...	...	...	...	...	...	...	...	*	*	*	...		...	...	...	...	...	...	...
E	...	...	...	...	...	...	...	...	...	...	...	...	...	...	...	*	...	...	...	...	...	...	...	...	...	...	...	...		...	*	...	...	...	...	...
F	...	...	...	...	...	...	...	...	...	...	...	...	...	...	...	...	...	...	...	...	...	...	...	...	...	...	...	...		*	...	...	...	...	...	...

The Fulicariæ include the following families :—

Sternum with two notches on each side of the posterior margin. No lateral bare tracts on neck.

Otididæ.—No hallux ; no oil-gland.

Opisthocomidæ.—Hallux large ; oil-gland tufted.

Sternum with one notch on each side of the posterior margin.

Cariamidæ.—No lateral bare tracts on neck ; oil-gland nude.

Heliornithidæ.—No lateral bare tracts on neck ; oil-gland tufted.

Rallidæ.—Long lateral bare tracts on neck ; oil-gland tufted.

Sternum with no notch on posterior margin.

Psophiidæ.—Long lateral bare tracts on neck.

Suborder 30. PYGOPODES.

The Grebes and the Divers may be diagnosed as follows:—

A. Cnemial process of tibia produced forwards to a remarkable degree. **B**. Posterior processes of the ilium approximated to such an extent that the sacrum is almost entirely concealed. **C**. Spinal feather-tract not defined on the neck. **D**. Palate schizognathous.

It is probable that the combination of characters **A** and **B** is diagnostic, and that the character **B** is unique.

	1. Passeres.	2. Eurylæmi.	3. Trochili.	4. Scansores.	5. Upupæ.	6. Trogones.	7. Columbæ.	8. Musophagi.	9. Cuculi.	10. Serpentarii.	11. Accipitres.	12. Striges.	13. Psittaci.	14. Halcyones.	15. Coraciæ.	16. Bucerotes.	17. Cathartes.	18. Steganopodes.	19. Herodiones.	20. Plataleæ.	21. Phœnicopteri.	22. Anseres.	23. Palamedeæ.	24. Tubinares.	25. Impennes.	26. Gaviæ.	27. Limicolæ.	28. Grallæ.	29. Fulicariæ.	30. Pygopodes.	31. Gallinæ.	32. Crypturi.	33. Apteryges.	34. Rheæ.	35. Casuarii.	36. Struthiones.
A	*	*	*	*	*	*	*	*	*	*	*	*	*	*	*	*	*	*	*	*	*	*	*	*	*	...	*	*	*		*	*	*	*	*	*
B	*	*	*	*	*	*	*	*	*	*	*	*	*	*	*	*	*	*	*	*	*	*	*	*	*	*	*	*	*		*	*	*	*	*	*
C	*	*	*	*	*	*	*	*	*	*	*	*	*	*	*	...	...	...	...	...	...	...	...	*	...	*	...	...	...		*	*	...	...	...	...
D	*	*	...	...	*	...	...	*	*	*	*	*	*	*	...	*	*	*	*	*	*	*	*	...	...	...	...	...	...		...	*	*	*	*	*

The Pygopodes consist of two families:—

Colymbidæ.—No ankylosed vertebræ in front of the ankylosed sacral vertebræ; median xiphoid process of sternum projecting behind the lateral processes. Ambiens and femoro-caudal muscles present. Number of cervical vertebræ 14 or 15.

Podicipidæ.—Ankylosed sacral vertebræ preceded by a free vertebra, in front of which are four ankylosed dorsal vertebræ; median xiphoid process of sternum abruptly truncated, so that the lateral processes extend behind it. Ambiens and femoro-caudal muscles absent. Number of cervical vertebræ 17 to 21.

Suborder 31. GALLINÆ.

The Pheasants and their allies may be diagnosed as follows:—

A. Basipterygoid processes articulating with the pterygoids as near the palatines as possible. **B.** Episternal process perforated to receive the feet of the coracoids.

This diagnosis may be supplemented by many other characters:—

C. Young born covered with down, and able to run in a few hours. **D.** Nasals holorhinal.

	1. Passeres.	2. Eurylæmi.	3. Trochili.	4. Scansores.	5. Upupæ.	6. Trogones.	7. Columbæ.	8. Musophagi.	9. Cuculi.	10. Serpentarii.	11. Accipitres.	12. Striges.	13. Psittaci.	14. Halcyones.	15. Coraciæ.	16. Bucerotes.	17. Cathartes.	18. Steganopodes.
A	*	*	*	*	*	*	*	*	*	*	*	*	*	*	*	*	*	*
B	*	*	*	*	...	*	*	*	*	*	*	*	*	*	...	...	*	*
C	*	*	*	*	*	*	*	*	*	*	*	*	*	*	*	*	*	*
D	...	...	...	...	...	...	*	...	...	...	...	...	...	...	...	...	...	...

	19. Herodiones.	20. Plataleæ.	21. Phœnicopteri.	22. Anseres.	23. Palamedeæ.	24. Tubinares.	25. Impennes.	26. Gaviæ.	27. Limicolæ.	28. Grallæ.	29. Fulicariæ.	30. Pygopodes.	31. Gallinæ.	32. Crypturi.	33. Apteryges.	34. Rheæ.	35. Casuarii.	36. Struthiones.
A	*	*	*	...	*	*	*	*	*	*	*	*		*	*	*	*	*
B	*	*	*	*	*	*	*	*	*	*	*	*		*	*	*	*	*
C	*	*	...	...	...	*	*	...	...	...	...	...		...	...	...	...	...
D	...	*	...	...	...	...	...	...	*	*	...	...		*	...	...	...	...

The Gallinæ include the following families:—Phasianidæ, Cracidæ, and Megapodiidæ.

Phasianidæ.—Oil-gland tufted, or absent; inner notch of sternum more than half the length of the whole sternum; hallux raised above the level of the front toes, and its basal phalanx shorter than that of the third toe.

Cracidæ.—Oil-gland tufted; inner notch of sternum less than half length of sternum; hallux on same level as other toes, and its basal phalanx as long as that of third toe.

Megapodiidæ.—Oil-gland nude; sternum more than twice the length of its inner notch; hallux on same level as other toes, and its basal phalanx as long as that of the third toe.

Suborder 32. CRYPTURI.

The Tinamous are perfectly unique in the form of the sternum, which has a narrow median xiphoid process to support the keel, and on each side a still narrower xiphoid process, the three processes occupying four fifths of its entire length.

They may otherwise be diagnosed in various ways :—

A. The keel of the sternum is well developed. **B.** The cartilage which connects the ilium with the ischium behind the acetabulum is not ossified.

This diagnosis is complete, as is also the following :—

C. The vomer coalesces with the maxillo-palatines in front, and with the pterygoids and palatines behind. **D.** The feather-tracts are well differentiated from the bare tracts both on the upper and under parts.

	1. Passeres.	2. Eurylæmi.	3. Trochili.	4. Scansores.	5. Upupæ.	6. Trogones.	7. Columbæ.	8. Musophagi.	9. Cuculi.	10. Serpentarii.	11. Accipitres.	12. Striges.	13. Psittaci.	14. Halcyones.	15. Coraciæ.	16. Bucerotes.	17. Cathartes.	18. Steganopodes.	19. Herodiones.	20. Plataleæ.	21. Phœnicopteri.	22. Anseres.	23. Palamedeæ.	24. Tubinares.	25. Impennes.	26. Gaviæ.	27. Limicolæ.	28. Grallæ.	29. Fulicariæ.	30. Pygopodes.	31. Gallinæ.	32. Crypturi.	33. Apteryges.	34. Rheæ.	35. Casuarii.	36. Struthiones.
A	...	...	...	...	...	...	...	...	...	...	...	...	...	...	...	...	...	...	...	...	...	...	...	...	...	...	...	...	...	...	...		*	*	*	*
B	*	*	*	*	*	*	*	*	*	*	*	*	*	*	*	*	*	*	*	*	*	*	*	*	*	*	*	*	*	*	*		...	...	...	...
C	*	*	*	*	*	*	*	*	*	*	*	*	*	*	*	*	*	*	*	*	*	*	*	*	*	*	*	*	*	*	*		...	...	...	...
D	...	...	...	...	...	...	...	...	...	...	...	...	...	...	...	...	...	...	...	...	...	...	*	...	...	...	...	...	...	...	...		*	*	*	*

The Tinamous appear to form a link between the Gallinæ and the Apteryges. The sternum is quite unique, but it approaches nearest to that of the Gallinæ, and is as far as possible from that of the Ratitæ. In their pterylosis the Tinamous are Galline, and not in the least Struthionine. On the other hand, the pelvis and the palate have certain characteristics which are only to be found in the Struthioniformes.

The inference to be drawn from these facts appears to be that the Struthioniformes are not so widely separated from the Galliformes as some ornithologists suppose them to be. If these two subclasses were thrown together, the diagnosis "Young born covered with down or feathers; palate not desmognathous; front plantar not leading to hallux," would cover them both.

Subclass STRUTHIONIFORMES.

The Struthioniformes combine two characters not known to be combined in any other birds :—

A. The keel of the sternum has become obsolete. **B.** Basipterygoid processes very large, and placed on the basisphenoid rather than on its rostrum.

This diagnosis may be strengthened by many other characters :—

C. Oil-gland absent. **D.** Plumage of neck continuous.

	1. Passeres.	2. Eurylæmi.	3. Trochili.	4. Scansores.	5. Upupæ.	6. Trogones.	7. Columbæ.	8. Musophagi.	9. Cuculi.	10. Serpentarii.	11. Accipitres.	12. Striges.	13. Psittaci.	14. Halcyones.	15. Coraciæ.	16. Bucerotes.	17. Cathartes.	18. Steganopodes.	19. Herodiones.	20. Plataleæ.	21. Phœnicopteri.	22. Anseres.	23. Palamedeæ.	24. Tubinares.	25. Impennes.	26. Gaviæ.	27. Limicolæ.	28. Grallæ.	29. Fulicariæ.	30. Pygopodes.	31. Gallinæ.	32. Crypturi.	33. Apteryges.	34. Rheæ.	35. Casuarii.	36. Struthiones.
A	*	*	*	*	*	*	*	*	*	*	*	*	...	*	*	*	*	*	*	*	*	*	*	*	*	*	*	*	...	*	*	*				
B	*	*	*	*	*	...	...	*	*	...	*	...	*	*	...	*	...	*	*	*	*	...	...	...	*	*	...	...	*	*	...	...				
C	*	*	*	*	*	*	...	*	*	*	*	*	...	*	...	*	*	*	*	*	*	*	*	*	*	*	*	*	...	*	...	*				
D	*	*	*	*	*	*	*	*	*	*	*	*	*	*	*	...	...	...	*	...	...	...	...	*	...	*	...	...	...	...	*	*				

There has been much diversity of opinion as to the precise boundaries of the Ratitæ. The widest scheme places the Struthiones, the Casuarii, the Rheæ, the Apteryges, and the Crypturi together under the name of Dromæognathæ (Stejneger, Stand. Nat. Hist. iv. p. 31), because they are all supposed to have the ilium posteriorly separated from the ischium, and are assumed to resemble each other in the details of their palates. A more restricted scheme excludes the Crypturi (Huxley, Proc. Zool. Soc. 1867, p. 426), because the keel of their sternum is so fully developed; whilst the most restricted scheme excludes the Apteryges also (Fürbringer, Morph. und Syst. Vögel, p. 1467), which are removed with the Crypturi to the Alectorornithes. The great variations in the so-called dromæognathous palate, and the facts that in the Crypturi the ilium is posteriorly united to the ischium by cartilage, and that in many skeletons (presumably of old birds) of *Casuarius* they are united by bone, are probably sufficient grounds for rejecting the widest scheme.

Order APTERYGES.

Suborder 33. APTERYGES.

The Kiwis may be diagnosed as follows :—

A. The keel of the sternum has become obsolete. **B.** Basipterygoid processes present. **C.** Hallux present.

Other characters may be added :—

D. Cervical vertebræ not more than 16 in number.

	1. Passeres.	2. Eurylæmi.	3. Trochili.	4. Scansores.	5. Upupæ.	6. Trogones.	7. Columbæ.	8. Musophagi.	9. Cuculi.	10. Serpentarii.	11. Accipitres.	12. Striges.	13. Psittaci.	14. Halcyones.	15. Coraciæ.	16. Bucerotes.	17. Cathartes.	18. Steganopodes.	19. Herodiones.	20. Plataleæ.	21. Phœnicopteri.	22. Anseres.	23. Palamedeæ.	24. Tubinares.	25. Impennes.	26. Gaviæ.	27. Limicolæ.	28. Grallæ.	29. Fulicariæ.	30. Pygopodes.	31. Gallinæ.	32. Crypturi.	33. Apteryges.	34. Rheæ.	35. Casuarii.	36. Struthiones.
A	*	*	*	*	*	*	*	*	*	*	*	*	...	*	*	*	*	*	*	*	*	*	*	*	*	*	*	*	...	*	*	*		...	...	...
B	*	*	*	*	*	...	...	*	*	...	*	...	*	*	...	*	...	*	*	*	*	...	...	...	*	*	...	...	*	*	...	...		...	...	...
C	...	...	...	...	...	...	...	...	...	...	...	...	...	...	...	...	...	...	...	...	...	...	...	...	...	...	...	...	...	...	...	...		*	*	*
D	...	...	...	...	...	...	...	...	...	...	...	...	...	...	...	...	...	...	*	*	*	...	*	...	...	...	...	...	...	...	...	...		...	*	*

In all the Struthioniformes the palatines articulate with the pterygoids, and not with the rostrum of the basisphenoid. In the arrangement of their palatine bones the Apteryges do not differ much from the Casuarii.

Order RATITÆ.

The Ratitæ combine three characters which are not known to occur together in any other birds:—

A. The keel of the sternum has become obsolete. **B.** Basipterygoid processes very large, and placed on the basisphenoid rather than on its rostrum. **C.** Hallux absent.

	1. Passeres.	2. Eurylæmi.	3. Trochili.	4. Scansores.	5. Upupæ.	6. Trogones.	7. Columbæ.	8. Musophagi.	9. Cuculi.	10. Serpentarii.	11. Accipitres.	12. Striges.	13. Psittaci.	14. Halcyones.	15. Coraciæ.	16. Bucerotes.	17. Cathartes.	18. Steganopodes.
A	*	*	*	*	*	*	*	*	*	*	*	*	...	*	*	*	*	*
B	*	*	*	*	*	...	...	*	*	...	*	...	*	*	...	*	...	*
C	*	*	*	...	*	*	*	*	*	*	*	*	*	*	*	*	*	*

	19. Herodiones.	20. Plataleæ.	21. Phœnicopteri.	22. Anseres.	23. Palamedeæ.	24. Tubinares.	25. Impennes.	26. Gaviæ.	27. Limicolæ.	28. Grallæ.	29. Fulicariæ.	30. Pygopodes.	31. Gallinæ.	32. Crypturi.	33. Apteryges.	34. Rheæ.	35. Casuarii.	36. Struthiones.
A	*	*	*	*	*	*	*	*	*	*	...	*	*	*	...			
B	*	*	*	*	...	...	*	*	...	...	*	*	*	...	...			
C	*	*	...	*	*	...	*	...	...	...	...	*	*	...	*			

Fürbringer regards the Apteryges as more nearly allied to the Crypturi and the Fulicariæ than to the Ratitæ; but it is not easy to discover any valid arguments for this conclusion.

Suborder 34. RHEÆ.

The Rheas may be diagnosed as follows :—

A. The keel of the sternum has become obsolete. **B.** The length of the humerus is more than the combined length of six dorsal vertebræ. **C.** Hallux absent ; second, third, and fourth digits present, and directed forwards. **D.** Basipterygoid processes present, and articulating with the pterygoids as near the quadrates as possible.

	1. Passeres.	2. Eurylæmi.	3. Trochili.	4. Scansores.	5. Upupæ.	6. Trogones.	7. Columbæ.	8. Musophagi.	9. Cuculi.	10. Serpentarii.	11. Accipitres.	12. Striges.	13. Psittaci.	14. Halcyones.	15. Coraciæ.	16. Bucerotes.	17. Cathartes.	18. Steganopodes.	19. Herodiones.	20. Plataleæ.	21. Phœnicopteri.	22. Anseres.	23. Palamedeæ.	24. Tubinares.	25. Impennes.	26. Gaviæ.	27. Limicolæ.	28. Grallæ.	29. Fulicariæ.	30. Pygopodes.	31. Gallinæ.	32. Crypturi.	33. Apteryges.	34. Rheæ.	35. Casuarii.	36. Struthiones.
A	*	*	*	*	*	*	*	*	*	*	*	*	...	*	*	*	*	*	*	*	*	*	*	*	*	*	*	*	...	*	*	*	...		...	...
B	...	...	*	...	...	...	...	...	...	...	...	...	...	...	...	...	...	...	...	...	...	...	...	...	*	...	...	...	...	...	...	...	*		*	...
C	*	*	*	*	*	*	*	*	*	*	*	*	*	*	*	*	*	*	*	*	...	*	*	...	*	...	...	...	...	*	*	...	*		...	*
D	*	*	*	*	*	...	...	*	*	...	*	...	*	*	...	*	...	*	*	*	*	...	...	...	*	*	...	...	*	*	*	...	...		...	...

The Rheæ appear to differ from all the other Struthioniformes in having the maxillo-palatines free from the vomer.

Suborder 35. CASUARII.

The Emus and the Cassowaries may be diagnosed as follows :—

A. The keel of the sternum has become obsolete. **B**. The vomer coalesces with the maxillo-palatines. **C**. Length of humerus less than the combined length of three dorsal vertebræ.

This diagnosis may be strengthened by the addition of other characters :—

D. Hallux absent; second, third, and fourth digits present, and directed forwards.

	1. Passeres.	2. Eurylæmi.	3. Trochili.	4. Scansores.	5. Upupæ.	6. Trogones.	7. Columbæ.	8. Musophagi.	9. Cuculi.	10. Serpentarii.	11. Accipitres.	12. Striges.	13. Psittaci.	14. Halcyones.	15. Coraciæ.	16. Bucerotes.	17. Cathartes.	18. Steganopodes.	19. Herodiones.	20. Plataleæ.	21. Phœnicopteri.	22. Anseres.	23. Palamedeæ.	24. Tubinares.	25. Impennes.	26. Gaviæ.	27. Limicolæ.	28. Grallæ.	29. Fulicariæ.	30. Pygopodes.	31. Gallinæ.	32. Crypturi.	33. Apteryges.	34. Rheæ.	35. Casuarii.	36. Struthiones.
A	*	*	*	*	*	*	*	*	*	*	*	*	...	*	*	*	*	*	*	*	*	*	*	*	*	*	*	*	...	*	*	*	...	...		...
B	*	*	*	*	*	*	*	*	*	*	*	*	*	*	*	*	*	*	*	*	*	*	*	*	*	*	*	*	*	*	*	...	...	...		...
C	*	*	*	*	*	*	*	*	*	*	*	*	*	*	*	*	*	*	*	*	*	*	*	*	*	*	*	*	*	*	*	*	*	*		*
D	*	*	*	*	*	*	*	*	*	*	*	*	*	*	*	*	*	*	*	*	...	*	*	...	*	...	...	...	...	*	*	...	*	...		*

The Casuarii differ from the Rheæ in the arrangement of the palatine bones; the vomer articulates with the maxillo-palatines in front. They also differ from the Struthiones in this character: the vomer articulates both with the palatines and with the pterygoids behind.

Suborder 36. STRUTHIONES.

The Ostriches may be diagnosed as follows:—

A. The keel of the sternum has become obsolete. **B.** Second digit lost.

To this diagnosis many other characters may be added:—

C. The length of the humerus is more than the combined length of six dorsal vertebræ. **D.** Cervical vertebræ more than 19 in number.

	1. Passeres.	2. Eurylæmi.	3. Trochili.	4. Scansores.	5. Upupæ.	6. Trogones.	7. Columbæ.	8. Musophagi.	9. Cuculi.	10. Serpentarii.	11. Accipitres.	12. Striges.	13. Psittaci.	14. Halcyones.	15. Coraciæ.	16. Bucerotes.	17. Cathartes.	18. Steganopodes.
A	*	*	*	*	*	*	*	*	*	*	*	*	...	*	*	*	*	*
B	*	*	*	*	*	*	*	*	*	*	*	*	*	...	*	*	*	*
C	...	...	*	...	...	...	...	...	...	...	...	...	...	...	...	...	...	...
D	*	*	*	*	*	*	*	*	*	*	*	*	*	*	*	*	*	...

	19. Herodiones.	20. Plataleæ.	21. Phœnicopteri.	22. Anseres.	23. Palamedeæ.	24. Tubinares.	25. Impennes.	26. Gaviæ.	27. Limicolæ.	28. Grallæ.	29. Fulicariæ.	30. Pygopodes.	31. Gallinæ.	32. Crypturi.	33. Apteryges.	34. Rheæ.	35. Casuarii.	36. Struthiones.
A	*	*	*	*	*	*	*	*	*	*	...	*	*	*	...	...	...	
B	*	*	*	*	*	*	*	*	*	*	*	*	*	*	*	*	*	
C	...	...	...	...	...	...	*	...	...	...	...	...	...	...	*	...	*	
D	*	*	*	...	*	*	*	*	*	...	*	...	*	*	*	*	...	

The arrangement of the palatal bones in the Struthiones is unique; the vomer is not prolonged backwards to articulate either with the palatines or the pterygoids.

INDEX.

THE END.

PRINTED BY TAYLOR AND FRANCIS, RED LION COURT, FLEET STREET.

Made in the USA
Monee, IL
15 November 2025

34945807R00039